Master the Language of the Universe

Practical Math. Made Simple. For Adults.

By Matthew Canning

influxa

Between the ages of five and seventeen, my primary responsibility in life was to learn. The sum of human knowledge and ingenuity was handed to me, but it wasn't until many years later that I came to appreciate the opportunity I had. I, like many, now find myself trying to jam learning in between the cracks and crevices of adult life.

I would give anything to have that chance again. This book is dedicated to the many teachers who wasted their time on me.

Contents

Introduction

My friends smiled and watched me from across the table. I looked up at them for a brief moment before shifting my attention back to the restaurant bill.

Okay, so $44.18. Tip would be twenty percent of forty-four, so...about...four times two, or five times two. Somewhere in between. Let's say nine. Okay. So forty-four plus nine...forty-five, forty-six, forty-seven...wait, well, forty-four plus <u>ten</u> is fifty-four, so one less than that. Fifty-three.

Right?

They were still watching. I shifted uncomfortably in my seat and tried to concentrate. The sounds of the restaurant swirled in my skull. Silverware clinked on plates. Women with shrill voices talked about their sisters' health problems. It all converged into an organic static that seemed to bypass my ears and crash directly into the frontal lobe of my brain.

Forty-three. Divided by three people. No, wait. <u>Fifty</u>-three. Right. So...fifty-three divided by three people. How do you even...?

As usual, patience eventually waned, and someone broke the silence from across the table: "About seventeen dollars, sixty-seven cents each." Everyone laughed, and with that, our traditional post-meal game came to a close.

When this traumatic event took place, I was an experienced management professional with two degrees in technical disciplines. Throughout my academic career, I consistently performed well in advanced math classes. I worked for years in the software engineering industry, where I used somewhat complex math on a near-daily basis. These facts supplied the setup to an ironic punch line: among my circle of friends, I was notoriously unable to perform quick mental calculations.

I never learned this skill, and always tried to imagine problems written out using the classical notations taught in school. I didn't understand how anyone could do even simple calculations quickly.

I set out to eliminate this shortcoming and discovered that solving problems like the one described can be easy. As I educated myself on the subject, my interest morphed into an obsession. Over time, I identified and solved issues one commonly faces when attempting to master practical math, and this effort eventually spawned the work you now hold in your hands.

We all need math. Although you're probably not going to have to speak with anyone about the Landau–Kolmogorov Inequality, you may have to look at a time sheet and quickly add up the hours worked. You may have to look at the menu at a cash-only café and figure out if you can afford a sandwich and a drink with the lone five-dollar bill in your wallet (accounting for sales tax). You will certainly have to perform value comparisons, such as figuring out if an item you'd like to purchase online is cheaper from *Vendor A* (higher price, free shipping) or *Vendor B* (slightly lower price, modest shipping fee). After your doctor tells you to ingest 1,300 mg of calcium each day, you may need to determine how many 280 mg capsules to take.

Within the tapestry of modern life, we encounter a constant flow of little practical math problems that must be done on the fly. They're often related to daily minutiae such as mileage, time zone differences, stock prices, conversions, or what test score you need to pull your grade into *B+* range. You need to use probability and "guess-timation," convert numbers to percentages, and perform other modest, mundane feats.

Most of us can do this type of simple, day-to-day math, but it either requires pencil and paper, a mobile phone calculator, or a few minutes of quiet thought to come up with the answer. By contrast, you probably read that last sentence—a complex structure involving nouns, adjectives, subjects, objects, pronouns, and punctuation—and understood it immediately.

This is because *you're comfortable with words*, the ideas they represent, and how they work together to tell a story or pose a question. Reading these sentences brings to mind a series of familiar associations. You don't concentrate on the sentence structure, the words themselves, or the letters that make up each word; rather, you absorb the context and information conveyed. English is a language you understand; the reason simple math scenarios seem more complicated than the English sentence is because you're not as well-versed in the language of mathematics.

In everyday life, too much effort is dedicated to the actual mathematical aspects of math problems, which distracts you from the context and meaning of the problems themselves. If you can become as comfortable and familiar with math as you are with English, the level of effort required decreases, and the "math" in the problem becomes less consequential.

In this book, you'll learn to become comfortable with math and stop being intimidated by day-to-day math problems. You'll tackle ways to deal with complicated math problems using tactics superior to those taught in school, allowing you to perform mathematical feats that appear to border on the supernatural.

With time, you'll learn to do some pretty advanced arithmetic entirely in your head. Before you can perform impressive feats of **Mental Calculation**, however, you must first learn how to approach different types of math problems using tools (a pencil and paper), and ease your way into **Mental Calculation** using only manageable, scalable methods that require very little carrying, borrowing, or peripheral complexity.

Learning to perform **Mental Calculation** isn't just useful in impressing nerdy romantic interests; unlocking mathematical abilities will broaden your capacity for logical reasoning, forging pathways in the mind otherwise left untapped. Math is the language of the universe, and to become fluent is to become universally wiser.

One could argue that daily life requires only the most basic addition, subtraction, multiplication, and division to be done without assistance. However, highly developed math skills can:

- Save you time. The more comfortable you are with simple math, the less time and effort go into solving problems and determining the best way to approach them.
- Make a great impression on others and increase your perceived reliability and decision-making value. Although it is not necessarily representative of overall intellect, people intuitively associate mathematical prowess with a high IQ.

- Give you an upper hand in negotiations and lessen the risk of getting "taken" or "scammed." In high-pressure sales pitch situations, you aren't often given time to pull out a calculator or crunch numbers properly. This is the point. Salespeople, armed with memorized figures, aim to catch consumers off-guard. Unfortunately, many feel as though it's a blow to the ego or a show of weakness to admit you're getting "lost in the numbers" and ask for a moment to think or write things down. The ability to quickly understand numbers and their consequences creates an impression of strength and control, which disarms aggressive sales tactics.

- Provide you with an invaluable tool for understanding your world. What's the better deal: 8 oz. of hot sauce for $2.99 or 13 oz. for $3.99? Would you expect the average person to know the answer? Would you know the answer? How long would it take you to figure out? This type of **Value Comparison** isn't all that difficult once the problem is deconstructed. However, most people don't know how and were never taught to perform such calculations properly.

What Sets Master the Language of the Universe Apart

Many of the core methods in this book have been explored and documented numerous times by different people; these techniques have survived rigorous scrutiny, were determined to be the most effective options for achieving the desired results, and are presented in this text in easily understood lessons.

In addition, *Master the Language of the Universe* contains a wealth of unique and innovative tactics that can't be found anywhere else, including:

- *Master the Language of the Universe*-specific multiplication and division **Shortcuts**.
- The concept of **Pre-Scanning** problems before approaching them to identify potential trouble spots.
- The **5-7 Approach** to **Mental Calculation**.
- A focus upon true fundamentals, including the ability to immediately solve any addition or subtraction problem involving two numbers under 20.
- Techniques pertaining to practical math skills like **Guestimation**, probability, percentages, **Sorting**, and **Value Comparison**.
- The **F2D Method**, a tool for quickly converting common fractions into precise decimals.
- A unique and modern approach to **Mental Calculation**.

The most important thing that sets *Master the Language of the Universe* apart is its "practicality first" approach. In most real-world cases, you won't need to solve problems down to the fifth decimal point. Rough answers are often perfectly adequate, so this book is split into two parts. You will first learn the skills required to produce fast, rough answers, and only then will you develop the tools needed to produce precise answers.

A Two-Step Process: First the Methods, Then the Mind

One of the greatest living mental calculators, Rudiger Gamm, is no ordinary man; he can speak backward in several languages and exhibits other fascinating mental abilities. Interestingly, Mr. Gamm's methods for performing speedy **Mental Calculation** are not strictly intuitive, like those of confirmed savants. Mr. Gamm claims to have been only average at arithmetic until his early twenties, when he began working diligently at developing and perfecting his own methods of speed calculation. Gamm's impressive ability is rooted in methods that can be learned and worked into the fabric of everyday life. Though clearly equipped with an atypical intellect, he is proof that mastering the right methods can turn anyone into a **Mental Calculation** machine.

However, you must crawl before you walk. Your journey from math novice to pro will be a two-step process: the first step is to learn the methods that make **Mental Calculation** possible (using a pencil and paper). Only after learning and reasonably mastering these methods should you move this process to your unassisted mind.

Note that "pencil and paper" doesn't mean you'll be scribbling *everything* down; even in the beginning, you will use a pencil to record the digits of your answers, but never to perform the computational steps. This applies to everything from the simplest fundamentals to the most advanced **Trachtenberg Math** (you'll learn about that soon). That's the beauty of this book; whenever possible, methods were chosen that don't require intermediate work to be written down. You'll understand why this is so as you move forward.

Let's Begin

You'll start with the types of problems most likely to crop up in daily life: small-number addition, subtraction, multiplication, and division. You'll work your way up to more advanced math skills as you become more comfortable with numbers and mathematics in general.

Section 1: Fundamental Mathematical Concepts

Many of us haven't touched upon remotely challenging math since late high school, and those "lucky enough" to use math in professional life often find themselves relying upon tools or laborious standardized methods learned years ago. Never questioning. Never challenging. Never asking the questions, "Is this method accomplishing the intended goal in a complete and satisfactory manner?" "Is there an entirely different method that can be explored?" or "Could my current method be executed in a simpler, faster, or more efficient manner?"

Re-learning math fundamentals (the math you *should* have learned in school) in a fresh, unique way will make you vastly more comfortable with numbers and prime you for more challenging feats ahead. Here you'll learn a series of methods for speeding up your ability to do common, simple math problems.

You shouldn't struggle to add or subtract two small numbers any more than you should struggle to read this sentence, and it's my mission in this section to eliminate this struggle. You may be saying, "Hey, I'm not an idiot! I already know how to add and subtract," but please continue on with an open mind. Resist the temptation to skip this section, as it's a vitally important foundation for the rest of the lessons.

The section begins with fundamental addition and subtraction skills, and then moves into multiplication and division.

Lesson 1: Adding Two Small Numbers

Oh, Addition! How thou hast scorned me.

If you're anything like I used to be, you'd be embarrassed to admit how bad you are at addition. It's okay; you're among friends.

Most people are never formally taught to solve addition or subtraction problems quickly in the mind. The fundamentals you'll learn here form the foundation of the methods used by the world's foremost mental calculators to perform addition upon two small numbers (numbers under 9,999).

Years ago, Jakow Trachtenberg (a man you'll soon know well) devised an incredibly useful method for adding three or more numbers together. Curiously, in all of his teachings, Trachtenberg foregoes any mention of a method for adding two numbers together. In fact, no notable historic mathematical figure has left us with useful advice for adding small numbers together beyond the classical methods taught in school.

The reason? The sheer simplicity of small addition problems doesn't leave much room for innovation. There's no "magic bullet."

Lazy, modern humans like us, spoiled from lifetimes of calculator use, can benefit from learning to solve small two-number problems quickly, as you probably encounter these types of problems fairly often in your daily life.

Knowing Your "Under Twenty" Tables

The best way to become more comfortable with addition and subtraction is by familiarizing yourself with the individual relationships between all numbers under twenty. When you memorized your multiplication tables, you could instantly recite "five times six is thirty" with no thought given to the calculation. In the same way, you should be able to instantly answer "what is thirteen minus eight?" Sure, you knew the answer, but how long did it take? Was it as quick as "ten minus one" or "four minus two?"

This is the first step of your journey: instant knowledge of every relationship (addition and subtraction) under twenty. Instant. Reactionary. Immediate. Get it?

This rule applies to "seventeen minus eight," "six plus nine," "four plus seventeen," and all other combinations that aren't quite as clean and pretty as "ten plus five." The universe isn't metric, and it doesn't play favorites for even numbers over odd or small numbers over large; you have to play the game on its terms.

$19 - 11.$

Again, I'm sure you know the answer. But was your reply instant? Take some time right now and memorize the following combinations:

Addition

+	1	2	3	4	5	6	7	8	9	10	11	12	13	14	15	16	17	18	19	20
1	2	3	4	5	6	7	8	9	10	11	12	13	14	15	16	17	18	19	20	21
2	3	4	5	6	7	8	9	10	11	12	13	14	15	16	17	18	19	20	21	
3	4	5	6	7	8	9	10	11	12	13	14	15	16	17	18	19	20	21		
4	5	6	7	8	9	10	11	12	13	14	15	16	17	18	19	20	21			
5	6	7	8	9	10	11	12	13	14	15	16	17	18	19	20	21				
6	7	8	9	10	11	12	13	14	15	16	17	18	19	20	21					
7	8	9	10	11	12	13	14	15	16	17	18	19	20	21						
8	9	10	11	12	13	14	15	16	17	18	19	20	21							
9	10	11	12	13	14	15	16	17	18	19	20	21								
10	11	12	13	14	15	16	17	18	19	20	21									
11	12	13	14	15	16	17	18	19	20	21										
12	13	14	15	16	17	18	19	20	21											
13	14	15	16	17	18	19	20	21												
14	15	16	17	18	19	20	21													
15	16	17	18	19	20	21														
16	17	18	19	20	21															
17	18	19	20	21																
18	19	20	21																	
19	20	21																		
20	21																			

Subtraction

1	2	3	4	5	6	7	8	9	10	11	12	13	14	15	16	17	18	19	20	-
0	1	2	3	4	5	6	7	8	9	10	11	12	13	14	15	16	17	18	19	**1**
	0	1	2	3	4	5	6	7	8	9	10	11	12	13	14	15	16	17	18	**2**
		0	1	2	3	4	5	6	7	8	9	10	11	12	13	14	15	16	17	**3**
			0	1	2	3	4	5	6	7	8	9	10	11	12	13	14	15	16	**4**
				0	1	2	3	4	5	6	7	8	9	10	11	12	13	14	15	**5**
					0	1	2	3	4	5	6	7	8	9	10	11	12	13	14	**6**
						0	1	2	3	4	5	6	7	8	9	10	11	12	13	**7**
							0	1	2	3	4	5	6	7	8	9	10	11	12	**8**
								0	1	2	3	4	5	6	7	8	9	10	11	**9**
									0	1	2	3	4	5	6	7	8	9	10	**10**
										0	1	2	3	4	5	6	7	8	9	**11**
											0	1	2	3	4	5	6	7	8	**12**
												0	1	2	3	4	5	6	7	**13**
													0	1	2	3	4	5	6	**14**
														0	1	2	3	4	5	**15**
															0	1	2	3	4	**16**
																0	1	2	3	**17**
																	0	1	2	**18**
																		0	1	**19**
																			0	**20**

This may take you a few days. Like anything in this book, the speed of your progress is far less important than skill development. After spending some time practicing these, you should be able to respond to 6 + 8 ("14"), 17 − 9 ("8"), or 12 + 9 ("21") immediately. Note that I used the word "respond," as opposed to "calculate;" there should be no *calculation* involved, much as you shouldn't have to think of the nouns, verbs, and adjectives in this sentence. Even if you're generally good with basic math like this, there are probably a few combinations that force you to think for a moment or two. This exercise was designed to replace any such hesitation with confidence.

Intimacy with these relationships (especially relationships between two numbers whose addition or subtraction cross a "tens-place line," such as "eight plus seven") is crucial for advancement using the strategies you'll learn next. By knowing these combinations inside out, you can quickly project this knowledge onto more complex relationships. Consider, for example, "112 – 8." If you know immediately that twelve minus eight equals four, all that's left is to do is retain the extra hundred throughout the process: 12 – 8 = 4, so $\underline{1}$12 – 8 = $\underline{1}$04.

Using Familiar Numbers

Familiar Numbers are numbers that are easy to work with because of your familiarity with metric increments. For some people, this means numbers that are multiples of five (5, 10, 15, 20...), but for most, it's much faster to use multiples of ten (10, 20, 30, 40...). I certainly fall into the latter category.

Once you've mastered the **Under Twenty Tables**, you can add any two small (< 9,999) numbers relatively easily using **Familiar Numbers**. Put simply, you'll convert the problem's numbers to **Familiar Numbers** to make them easier to add.

You're going to learn two ways to use **Familiar Numbers** with addition, and you can ultimately adopt whichever one you prefer.

Single-Offset Addition

The difference between a number (for example, 84) and a nearby **Familiar Number** (in this case, 80) is called the **Offset** (here, it would be 4). The first method you'll learn, **Single-Offset Addition**, only requires the use of a single **Offset** (hence the name).

To illustrate the concept **Single-Offset Addition**, consider the problem 142 + 29. These are *not* **Familiar Numbers** (as neither is a multiple of ten), and so they do not add together quickly or smoothly. The first step is to convert one of the two to a **Familiar Number** (a multiple of ten). It's usually easier to do this with the smaller number (29, in this example). You should quickly figure out that 29's closest multiple of ten is 30. Add this **Familiar Number** (30) to the larger number (142) and you'll get 172.

Hold this number in your head for a moment. Now, simply adjust the answer (172 so far) to account for the rounding you did while **Familiarizing** the number 29. Since 29 rounded up to 30, you're left with an **Offset** of (negative) 1.

This is where this method can seem complex for beginners; when **Familiarizing** a number by rounding *up* (29 *rounded up* by 1 is 30), your **Offset** will be negative. If you *round down* when **Familiarizing** a number (for example, 63 *rounded down* by 3 equals 60), your **Offset** will be positive.

This phenomenon is illustrated by finishing the example problem; if you take the **Offset** of (negative) 1 and add it to the original answer (172), you arrive at the final answer, 171.

Though this may sound involved, it really comes down to four simple steps:

1. Convert the smaller of the problem's numbers to a **Familiar Number**.

2. Add the larger number to the new **Familiar Number**.
3. Figure out the **Offset**.
4. Account for the **Offset**.

Tip: When performing **Single-Offset Addition**, it can be difficult to remember whether to add or subtract the **Offset**. Instead of thinking of it numerically—in terms of negatives and positives— you can use your imagination to create a visual or physical association. In the example above (142 + 29), while performing the first step (converting 29 to 30), you can imagine an upward motion to symbolize going "up" from 29 to 30. The term "upward motion" is purposely vague; "upward motion" can mean almost anything you'd like. Focusing on an inhalation (an "upward" breath) can be a great tool for this, if you think of the cadence of the problem as that of a complete breath. Once a breath has been taken (at the start of a problem), it must be released (at the end). Once you "breathe in," get your pre-**Offset** answer (step 2), and figure out the **Offset** (step 3), imagine a shift back "down" when accounting for the **Offset** (coming "down" by 1), like an exhalation.

In the same way, had you rounded down in order to solve a problem (like you would with 166 + 22), you could first imagine the downward motion/exhalation while converting 22 to 20, and then imagine an upward motion/inhalation when adding the **Offset** of (positive) 2.

166 + 20 + 2 = 188.

Perform the following calculations using **Single-Offset Addition**:

1. 121 + 63
2. 584 + 15
3. 66 + 711
4. 320 + 83

Answers (and explanations):

1. 63 becomes 60. 121 + 60 = 181. The offset is 3. 180 + 4 = 184.
2. 15 becomes 10. 584 + 10 = 594. The offset is 5. 594 + 5 = 599.
3. 66 becomes 70. 711 + 70 = 781. The offset is −6. 781 − 6 = 777.
4. 83 becomes 80. 320 + 80 = 400. The offset is 3. 400 + 3 = 403.

Dual-Offset Addition

The second option for addition using **Familiar Numbers** is called **Dual-Offset Addition**. Although marginally more complex, this can be faster if you're already proficient at adding small numbers (and you *should* be, if you did your homework from earlier).

To illustrate the concept of **Familiar Numbers** in the context of **Dual-Offset Addition**, consider again the problem 142 + 29.

Again, these are not **Familiar Numbers** (neither is a multiple of ten), and they do not add together smoothly. Using this method, the first step is to convert *both* to **Familiar Numbers**. The closest multiples of ten are 30 (for 29) and 140 (for 142). Add the two **Familiar Numbers** (30 + 140 = 170).

Next, adjust the answer to account for the rounding you did while **Familiarizing** the two numbers; 29 rounded up to 30, leaving you with an **Offset** of −1, and 142 rounded down to 140, leaving you with an **Offset** of +2.

If you take the −1 **Offset** from first number and add it to the +2 **Offset** from the second, you have a **Final Offset** of +1. Now add this **Final Offset** (+1) to the pre-**Offset** answer (170) to arrive at the final answer, 171.

It all comes down to five simple steps:

1. Convert both of the problem's individual numbers to **Familiar Numbers**.
2. Add the two Familiar Numbers.
3. Figure out the **Offset** of each of the numbers.
4. Add the two **Offsets** together to arrive at a **Final Offset**.
5. Account for the **Final Offset**.

Let's look at a slightly harder example: 2,368 + 1,114. In this example, you would:

1. Round 2,368 to 2,370 and round 1,114 to 1,110.
2. Add the rightmost three digits of the **Familiar Numbers** together: 370 plus 110 is 480. Next, add the thousands place digits of the **Familiar Numbers** together: 2,000 plus 1,000 equals 3,000. Add the sum of the thousands place digits to the sum of the rightmost three digits: 3,000 plus 480 is 3,480. Hold this number in your head.

3. Remember that you rounded *up* by 2 to **Familiarize** the first number (a negative **Offset**), and *down* by 4 to **Familiarize** the second number (a positive **Offset**).
4. Adding these two **Offsets** (-2 + 4) gives a **Final Offset** of +2.
5. Lastly, add the **Final Offset** (+2) to the pre-**Offset** answer (3,480) to arrive at 3,482.

You're not going to practice **Dual-Offset Addition** just yet, as multiple **Offsets** can be a bit much to keep track of. We'll first go over how to make this simpler.

Keeping It Simple: One-Way Rounding

Just now, you rounded one number *up* to the nearest **Familiar Number** (2,368 to 2,370), but rounded the other *down* (1,114 to 1,110). This method may work for those who are math-savvy to begin with; however, for those who are not, uniformly rounding *either* up or down for **Dual Offset Addition** is faster, safer, and easier to learn.

To demonstrate, take the 2,368 + 1,114 example again. 2,368 is much closer to 2,370 than it is to 2,360, so it's natural to want to round *up* from 2,368 to 2,370. The second number, 1,114, is closer to 1,110 than it is 1,120, which naturally causes you to want to round *down*. The **One-Way Rounding Rule** tells you to round *both numbers up* (2,370 + 1,120, **Final Offset** of −8) or *both numbers down* (2,360 + 1,110, **Final Offset** of 12).

In many cases, this means that you won't be rounding in the most efficient direction for one of the two numbers, but this sacrifice allows for a manageable degree of simplicity.

We can reinforce this concept by going back to the very first example (142 + 29). Even though 142 is closer to 140 (down) than it is to 150 and 29 is closer to 30 (up) than 20, either round *both up* (30 plus 150 equals 180, **Final Offset** of –9) or *both down* (20 plus 140 equals 160, **Final Offset** of 11).

You may ask, "So, which direction? Up or down?"

Ultimately, it doesn't really matter, but there is a way to determine which one makes the most sense given the specific numbers in question. If rounding either up or down results in the **Final Offset** exceeding (positive or negative) 9, choose the other direction, as it will be much easier to work with. You can see this in action by going back to earlier examples. Rounding 2,368 + 1,114 down gives you a **Final Offset** of 12, and rounding 142 + 29 down gives you a **Final Offset** of 11, so you should round *up* in these cases: rounding 2,368 + 1,114 up gives a **Final Offset** of –8, and rounding 142 + 29 up gives a **Final Offset** of –9.

There are a few odd cases to consider.

First, when dealing with two numbers that both end in "5," the **Offset** for each will be +/–5 (whether rounded up or down) and the **Final Offset** will equal +/–10, so the rounding direction is unimportant.

Second, in cases where one of the numbers is already a **Familiar Number**, the **Final Offset** can't exceed 9.

Finally, when you're adding two numbers with ones place digits that add up to 10 (such as 1 and 9 or 2 and 8), you'll end up with a

Final Offset of +/−10, and thus the rounding direction is unimportant in this case as well.

Let's do an exercise. Look at the following problems and decide which direction to round them so their **Final Offsets** equals 9 or less. Don't worry about solving the problems for now.

1. 617 + 814
2. 832 + 910
3. 239 + 612
4. 1,786 + 1,855
5. 781 + 69

Answers:

1. Up
2. Down
3. Up
4. Up
5. Does not matter (Unavoidable two-digit **Offset**)

In the context of daily life, the ability to perform addition using **Familiar Numbers** can be an incredibly useful tool. Imagine that it's 4:46 PM and something needs to come out of the oven in 28 minutes. Round up (inhale): 4:46 becomes 4:50 and 28 minutes becomes 30. Adding 30 minutes to 4:50 gives you 5:20, and the **Final Offset** is −6. Finally, you drop (exhale) the 6-minute **Final Offset** from the pre-**Offset** answer (5:20) to arrive at 5:14.

It sounds like quite a few steps, but with time and practice, it begins to come naturally. Most people who are performing quick mental addition approach problems using some variation of this method.

Complete the following computations in your head. Use **Dual-Offset Addition** and **One-Way Rounding**:

1. 813 + 19
2. 422 + 164
3. 928 + 27
4. 1,394 + 4,652

The answers are:

1. 832
2. 586
3. 955
4. 6,046

There are only two skills necessary to master addition using **Familiar Numbers**. The first is becoming fast at associating numbers with their closest rounded multiple of ten (this means hearing "117" and thinking "120"). The second is becoming fast at adding the two **Familiar** numbers; this is where instinctual knowledge of every numerical relationship below twenty comes in handy. With mastery of these two skills, further development becomes a simple matter of practice.

Don't Panic!

"If you can't perform this addition problem in twenty seconds, Mr. Bond, your lovely partner will be lowered into this pit of molten lead."

Admittedly, this scenario is unlikely to find its way into a *007* film anytime soon, but the anxiety it conveys probably feels familiar; after all, this scene is only marginally more stressful than, "Add these receipt items together while I glare at you, waiting for you to tell me the sum."

When performing addition—whether in your head or on paper—it's important not to panic. Often, before improving at the type of basic arithmetic explored in this section, people can be so self-conscious about how long it takes to add or subtract two numbers that they resort to desperately stabbing at the answer with no real direction.

Take a breath, worry less about what others think as you develop this skill, and take as long as you need to perform calculations properly. Know that the moments are ticking by much more slowly for you than they are for everyone else. Resist the urge to guess at any cost.

Quick Review

1. The first step toward achieving mathematical fluency is to become incredibly familiar with every possible addition and subtraction relationship involving numbers under twenty. The ability to produce these sums and differences in a confident,

instant, and instinctual manner better prepares you to perform more difficult addition and subtraction problems quickly.

2. Using **Familiar Numbers** (rounding numbers up or down to their nearest multiple of ten), you can tackle daunting addition or subtraction problems with ease.

3. **Single-Offset Addition** involves rounding one of a problem's two numbers (usually the smaller number) and accounting for the **Offset**, or the amount by which it was rounded.

4. **Dual-Offset Addition** involves rounding both of a problem's numbers and accounting for their **Offsets**, or the amounts by which they were rounded.

Lesson 2: Subtracting Small Numbers

You now know how to add two small numbers; now you'll learn how to perform subtraction upon such numbers.

The Two Methods of Subtraction

Like addition, subtraction involving small numbers can be done two ways. The first (**"Addition" Subtraction**) is a method you can find—in various forms—in texts dating back hundreds of years. This is suggested for those who struggle with quick **Mental Calculation** (so, most of us). The second (**Dual Offset Subtraction**) employs your good friends, **Familiar Numbers**, and though more complex, can produce faster answers if you're more math-savvy.

Addition Subtraction

Addition Subtraction is so named because it involves splitting a difficult subtraction problem into two smaller problems: one simpler subtraction problem and one addition problem. The brain is generally more comfortable performing addition than subtraction. In order to perform **Addition Subtraction**, use the following steps:

1. Round the smaller of the two numbers *up*, using the largest scale available to you. This means if the number has 3 digits, round it up to the nearest multiple of 100 (example: 642 becomes 700); if it has 2 digits, round it up to the nearest multiple of 10 (example: 87 becomes 90). The fact that you're rounding up in all cases may seem a bit odd (using this method, 444, 412 and 498 would all become 500, and 68, 61, 69 and 65 would all become 70), but—as you'll see—it makes sense to do so.
2. Subtract this rounded number from the larger number.
3. Calculate the smaller number's **Offset** (the distance between the rounded number and the original number).
4. Add the **Offset** to the outcome of Step 2.

Let's walk through an example: 618 − 281.

1.	281 becomes 300
2.	618 − 300 = 318
3.	Offset = 19
4.	318 + 19 = 337

Answer: 337.

Moving forward in this book, you'll see work performed in a box like the one above.

Let's walk through another: 876 − 192. Think about it carefully and walk yourself through each step.

1.	192 becomes 200
2.	876 − 200 = 676
3.	Offset = 8
4.	676 + 8 = 684

Answer: 684. If it took you more than a moment to perform the final calculation (676 + 8), please be sure that you've memorized your **Under Twenty Tables**.

Before learning about **Dual-Offset Subtraction**, do a few more problems using **Addition Subtraction**:

1. 942 − 189
2. 2,378 − 1,102
3. 4,019 − 198
4. 540 − 64
5. 388 − 192

Answers:

1. 753
2. 1,276
3. 3,821
4. 476
5. 196

It's common for everyday subtraction problems to involve two three-digit numbers; here's a shortcut to help you quickly calculate the **Offset** in such cases. In their book, *Secrets of Mental Math*, Arthur Benjamin and Michael Shermer point out that the **Offset** in these problems can be computed by subtracting the tens place digit of the rounded number from 9, and subtracting the ones place digit from 10. For example, if you were to subtract 158 from 374:

1. 158 becomes 200 (as taught).
2. 374 − 200 = 174, (as taught).
3. Use Benjamin and Sherman's tip. The smaller number was 158, so focus on the tens and ones places (the 58 from 1<u>58</u>): 9 − <u>5</u> = 4 and 10 − <u>8</u> = 2. Your **Offset** is 42.
4. 174 + 42 = 216 (using your preferred method of addition—**Single Offset** or **Dual Offset**).

Answer: 216.

This method involves a few steps, but they're all incredibly simple.

Take another look at the three problems from above that involved only three-digit numbers. This time, perform the calculations using Benjamin and Shermer's shortcut.

1. 942 − 189
2. 540 − 64
3. 388 − 192

Answers:

1. 753
2. 476
3. 196

Dual-Offset Subtraction

Subtraction involving small numbers can also be performed without addition using **Familiar Numbers**; simply perform the problem using **Familiar Numbers**, subtract the smaller number's **Offset** from the larger number's to arrive at a **Final Offset**, and then subtract the **Final Offset** from the pre-**Offset** answer.

As with addition, rounding both numbers in the same direction will minimize potential confusion when implementing this method. Let's revisit the same numbers you used when first learning to *add* **Familiar Numbers** (142 and 29); you will see how the same general method works for subtraction.

Rounding in the same direction (up, in this case), 142 becomes 150 and 29 becomes 30. So, you end up with 8 − 1, or 7 for your **Final Offset**. Note that, as with **Dual-Offset Addition**, there is no need to round numbers that end in 0.

So far, you've rounded both numbers and determined your **Final Offset**. Next, subtract 30 from 150 (which gives you a pre-**Offset** answer of 120), and subtract the **Final Offset** (7) from the pre-**Offset** answer to arrive at 113.

Just to set your mind at ease, I'll show you that this would have worked had you chosen to round down instead of up. 142 rounds down to 140 and 29 rounds down to 20. 140 − 20 = 120. Next, calculate the **Offsets**: −2 − (−9) = 7. Regardless of the rounding direction, subtracting 7 from 120 leaves you with 113.

Now that you've worked through an example, read through this explicit breakdown of the steps:

1. Round the two numbers in the same direction (though the chosen direction doesn't matter), to arrive at two **Familiar Numbers**.
2. Subtract the smaller **Familiar Number** from the larger one to arrive at the pre-**Offset** answer.
3. To determine the **Final Offset**, subtract the smaller number's **Offset** from the larger number's **Offset**.
4. Subtract the **Final Offset** from the pre-**Offset** answer.

Let's do another example: 54 – 28. When performing these calculations, write down only the problem and the solution; do all the intermediate steps in your head. Never write down any **Familiar Numbers, Offsets, Final Offsets** or pre-**Offset** answers.

Option 1, *rounding down*:

1.	54 becomes 50, and 28 becomes 20.
2.	50 – 20 = 30 (pre-**Offset** answer).
3.	–4 – (–8) = 4, so the **Final Offset** is 4.
4.	30 (pre-**Offset** answer) – 4 (**Final Offset**) = 26.

Option 2, *rounding up*, same conclusion:

1.	54 becomes 60, and 28 becomes 30.
2.	60 – 30 = 30 (pre-**Offset** answer).
3.	6 – 2 = 4, so the **Final Offset** is 4.
4.	30 (pre-**Offset** answer) – 4 (**Final Offset**) = 26.

Do the following computations, using **Dual-Offset Subtraction**:

1. 819 – 13
2. 422 – 164

3. 928 – 39
4. 4,394 – 1,652

The answers are:

1. 806
2. 258
3. 889
4. 2,742

Which of these two subtraction methods do you prefer? Ultimately, it's a matter of personal preference. I urge you to try both and see how you feel about each. Which is quicker? Which is less frustrating? Which seems more natural?

Quick Review

1. **Addition Subtraction** is a method of subtraction by which you split a difficult subtraction problem into two smaller problems: a simpler subtraction problem and an addition problem.
2. **Dual-Offset Subtraction** is a method of subtraction by which you round both numbers in the same direction, and then subtract the **Final Offset** from the pre-**Offset** answer.

Lesson 3: Addition and Subtraction— Larger Numbers

Wilson (who famously played Biff Tannen in *Back to the Future*) isn't alone, but there are ways to make all math problems—even larger, more daunting ones—much more manageable.

When it comes to larger numbers (five or more digits), you'll use different methods from those you just learned. To take on larger challenges while dealing simply with **Familiar Numbers** would exponentially increase the chance for error. This is due to the increased need to retain information in the short term, especially while you are still learning to control your concentration and memory.

Adding Slightly Larger Numbers

To solve problems that involve five or more digits, the best method is actually quite simple: add from left to right. Yes, I'm aware that this amounts to spitting in the face of your fourth-grade math teacher, but soon you'll be running circles around her.

At this point, remembering a problem's two numbers long enough to perform the required addition should still be difficult, so feel free to write your problems and solutions down; however, don't use a paper and pencil to perform the intermediate addition steps.

Get right into it and add two numbers together from left to right: 32,345 + 53,401

To do this, you must become familiar with the idea of an **Addition Group**. In an addition problem that involves two numbers, the digits that occupy each respective "place"—such as the *tens* place or *hundreds* place, for example—are together considered an **Addition Group**.

The following is your first **Addition Group**:

32,345 + 53,401

Second **Addition Group**:

32,345 + 53,401

Third **Addition Group**:

32,345 + 53,401

...and so on.

For visualization purposes, the two digits in each **Addition Group** would line up if you were adding them together using classical addition methods:

32,345
53,401

Moving from left to right, look at the first **Addition Group** (3 and 5) and add them together: you should know immediately that 3 + 5 = 8. Then, add the next two digits together (2 and 3): 2 + 3 = 5. So far, the answer begins with "eighty five thousand..."

Continue to add each remaining **Addition Group** (3 + 4 = 7; 4 + 0 = 4; and 5 + 1 = 6), until you arrive at the final answer: 85,746. This example presented a perfect scenario: Every **Addition Group**, when added, left you with a sum smaller than ten, so there was no need to carry. Carrying introduces complexity to this left-to-right method, but there's a simple way to make it manageable: Before settling upon the final sum for each individual **Addition Group**, look to its neighbor (the next **Addition Group** immediately to its right), and see if that **Addition Group's** sum is equal to or greater than nine.

Why do this?

When performing this type of addition, any **Addition Group** that adds up to ten or greater requires you to "carry" (to use the term you probably learned in school). If the sum of the neighboring **Addition Group** equals ten or greater, this "carry" becomes the responsibility of your current **Addition Group**. You also have to think about **Addition Groups** that add up to nine, because they have the *potential* to take on a "carry"—which we'll call an **Overflow**—from *its* neighbor (turning the nine to a ten), and in doing so, pass an **Overflow** on to your current **Addition Group.** Such an **Overflow** has the potential to be passed from **Addition Group** to **Addition Group**, all the way from right to left. You'll need to look ahead in this manner until you're sure you're risk-free and able to confidently announce your current **Addition Group's** sum.

When the neighboring **Addition Group** adds up to ten, you know for certain it will be your responsibility to address it. When it adds up to nine, we'll call this **Addition Group** an **Overflow Risk.**

This all sounds terribly complicated, but it truly isn't. Let's look at an example: 12,841 + 16,129.

This example is less ideal than the first, as there's some **Overflow** in the mix. In this example, working from left to right, you see that 1 + 1 = 2 (from 12,841 + 16,129). Before you enthusiastically announce "twenty...," first make sure that the *next* **Addition Group** (the 2 and 6 from 12,841 + 16,129) doesn't add up to 9 or greater. Here it doesn't, so you can finish saying "twenty..." with hearty confidence.

Next, look at the second **Addition Group,** 2 + 6 = 8 (from 12,841 + 16,129). Again, you must check the neighboring **Addition Group** (the one to its right) before continuing. In this case, the neighboring **Addition Group** (the 8 and 1 from 12,841 + 16,129) *does indeed add up to 9*, and so presents an **Overflow Risk.**

Because of this, you need to ask if you must increase your current **Addition Group's** sum by 1 in order to account for an **Overflow** (since your current **Addition Group** is 2 + 6 = 8, this would involve changing it from 8 to 9). The answer lies in the *next* **Addition Group,** 4 + 2 (from 12,841 + 16,129). This **Addition Group** does not add up to 10 or more,

36

and so it will not pass an **Overflow** to its left-hand neighbor, 8 + 1 = 9. Your current **Addition Group's** sum (1**2**,841 + 1**6**,129) is simply 8.

So far, you can confidently say, "twenty eight thousand..."

Next, add the 8 + 1 (from 12,**8**41 + 16,**1**29) and note that there's no overflow risk from the next **Addition Group** (4 + 2 < 9). You can thus conclude that the next part of the answer is "nine hundred..."

After this, add 4 + 2 (from 12,8**4**1 + 16,1**2**9), but note that the next **Addition Group** (9 + 1 from 12,84**1** + 16,12**9**) will certainly cause an **Overflow**, so adjust 4 + 2 to equal 7 ("seventy...").

Then add the final (rightmost) **Addition Group**: 9 + 1 = 10. Since you moved 10 over to the previous **Addition Group** in the **Overflow**, this ends up being 0.

The final answer is 28,970.

Let's review a few points:

- If the neighboring **Addition Group** adds up to eight or less, it has no chance of affecting the outcome of your current **Addition Group**.
- If it adds up to ten or more, it will *immediately* cause an **Overflow**, so increase your current **Addition Group's** sum by one.
- If it adds up to nine, it poses a *risk* of **Overflow**; check the sum of the *next* **Addition Group** (now *two* to the right of the current one). If *this* **Addition Group** adds up to ten or more, that **Overflow** will pass to its neighbor, and if that neighbor

adds up to nine, it will then pass the **Overflow** along, causing you to increase your current **Addition Group's** sum by one.

In short, keep an eye on all **Addition Groups** to the right of the current one that could feasibly affect it.

Let's do another quick one together. When adding 341 + 481, you add the first **Addition Group** (3 + 4 = 7), but hang onto the answer, because looking ahead reveals that the next **Addition Group** (4 + 8) adds up to 12 and will cause an **Overflow**. Increase the first **Addition Group's** sum from 7 to 8 and move on, shifting your attention to the second (middle) **Addition Group** (4 + 8). Note that this middle **Addition Group's** neighbor (1 + 1) poses no **Overflow Risk**, so simply take the sum (12), record the ones place (2) and ignore the tens place (1), because it has already been absorbed into the sum of the first **Addition Group**. The final **Addition Group's** sum is 2, so the answer is 822.

Do a small example on your own: 2,241 + 1,918. Read on once you've finished.

How did that go? When adding the first (leftmost) **Addition Group** (2 + 1), you should have immediately noticed that the neighboring **Addition Group** (2 + 9) created an **Overflow** and adjusted accordingly. You may have also noticed that the rightmost **Addition Group** adds up to nine; however, as there was no **Addition Group** to its right, it was at no risk of inheriting **Overflow**. Was your final answer 4,159?

Here's a "nightmare" scenario: 837,562 + 162,726. If there are multiple 9's in a row, the **Overflow** can be passed through quite a few

Addition Groups, like a game of mathematical "hot potato." Let's break this mess down.

Although seemingly innocuous, this problem is terrible. The leftmost **Addition Group** (8 + 1) adds up to 9, and so you need to look at the neighboring **Addition Group** (3 + 6) to see if it too adds up to 9. Damn, it does. Okay, how about the next one (7 + 2)? You've got to be kidding me! Okay, how about the next (7 + 5)? Finally, *four* places from the start, you find two numbers (5 + 7) that are greater than nine when added, causing a domino effect of **Overflow** to begin. This **Overflow** gets passed to 7 + 2, then 3 + 6, and finally to your current (and first) **Addition Group,** 8 + 1. This will cause the 8 + 1 **Addition** Group to equal 10, meaning your answer begins with "one million..." Not a pretty sight.

Next, move on to the next **Addition Group,** 3 + 6. You do the same thing, looking to the right until you find relief from the **Overflow,** which you find again at the fourth **Addition Group,** 5 + 7. This **Overflow** flows back to your current **Addition Group,** turning your 9 (from 3 + 6) to 10. Ignoring the tens place, this gives you a 0. Next, shift your attention to the third **Addition Group,** 7 + 2. Continue on your own.

This example serves to illustrate a worst-case scenario that you may encounter when dealing with this method, and how the difference between an infuriating problem and a simple problem can be subtle; if the fourth **Addition Group** (5 + 7) had not produced an **Overflow,** all four of the preceding **Addition Groups** could have simply been added.

With practice, you'll learn to intuitively recognize patterns when you first face problems; any numbers containing several "higher" individual digits (sixes or higher) run the risk of multiple **Overflows,** and those which contain many "lower digits" (fours or lower) are usually easier. You'll get plenty of practice with this method at the end of this section.

Subtracting Slightly Larger Numbers

Subtraction problems that involve at least one larger number (five or more digits) can be solved with a similar approach.

Moving from left to right, subtract each digit of the smaller number from its mate in the larger number. As you may have already guessed, these pairs of digits form **Subtraction Groups**.

With problems like 65,897 − 41,234, there is little thinking to do. In each **Subtraction Group**, the larger number's digit is greater than the corresponding digit in the smaller number; simply subtract each smaller digit from each larger digit.

1. 6 − 4 (from <u>6</u>5,897 − <u>4</u>1,234) = 2
2. 5 − 1 (from 6<u>5</u>,897 − 4<u>1</u>,234) = 4
3. 8 − 2 (from 65,<u>8</u>97 − 41,<u>2</u>34) = 6
4. 9 − 3 (from 65,8<u>9</u>7 − 41,2<u>3</u>4) = 6
5. 7 − 4 (from 65,89<u>7</u> − 41,23<u>4</u>) = 3

The answer is 24,663.

Of course, it's not always this easy. Problems like 48,801 − 21,624 will present some challenges. Let's work through this example and discuss how to approach each step.

Beginning from the far left, your first **Subtraction Group** contains 4 and 2 (from <u>4</u>8,801 – <u>2</u>1,624). No problem here. But before committing to a difference, you must look to the neighboring **Subtraction Group** (the one located immediately to its right), and ensure no borrowing has to take place. The next **Subtraction Group** is 8 – 1 (from 4<u>8</u>,801 – 2<u>1</u>,624), so you're in the clear. Knowing nothing will affect the first **Subtraction Group's** calculation (4 – 2), you can be sure that the first answer digit is 2.

Now you can work on the next **Subtraction Group**, which is 8 – 1 (from 4<u>8</u>,801 – 2<u>1</u>,624). Again, before committing to the difference (as of now, 7), first check the neighboring **Subtraction Group**. The neighboring **Subtraction Group** is 8 – 6 (from 48,<u>8</u>01 – 21,<u>6</u>24). This requires no borrowing, and so the second digit of the answer (8 – 1) *definitely* equals 7. So far you have "twenty-seven thousand..."

Moving on, look at the hundreds place **Subtraction Group**. 8 – 6 = 2. Simple enough; however, look over at the neighboring (tens place) **Subtraction Group**, 0 – 2 (from 48,8<u>0</u>1 – 21,6<u>2</u>4). The current **Subtraction Group** (the hundreds place) must be borrowed from. Decrease your current **Subtraction Group's** answer by 1. This means that 8 – 6 will equal 1 as opposed to 2.

Do you understand what you did? Whenever the neighboring **Subtraction Group** needs to borrow, you decrement your current **Subtraction Group's** answer by 1.

Your answer so far is "twenty-seven thousand, one hundred and..."

Next, look at the tens place (48,8<u>0</u>1 – 21,6<u>2</u>4). [1]0 – 2 equals 8, but before committing to the difference, you must again check the neighboring **Subtraction Group**. Here, it's the ones place **Group**: 1 – 4 (from 48,80**1** – 21,62**4**). This **Subtraction Group** also needs to borrow, so decrease your current **Subtraction Group**'s answer—the tens place answer—by 1. Now, [1]0 – 2 = 7, not 8.

The answer so far is "twenty-seven thousand, one hundred and seventy…"

Finish with the ones place **Subtraction Group**; [1]1 – 4 (from 48,80**1** – 21,62**4**) equals 7.

The final answer is 27,177.

This technique only becomes complex if any of the answer digits turn out to be a 0 that needs to be borrowed from. In such cases, decrement the 0 to 9, and do the same to the **Subtraction Group** to its left. To avoid this unpleasantness, when a **Subtraction Group's** difference turns out to equal 0, check both the neighboring **Subtraction Group** *and* the previous step's answer. Let's look at an example: 17,903 – 1,997.

Since we're subtracting a four digit number from a five digit number, we can take a little shortcut in the first step and subtract 1 from 17 (instead of 0 from 1 and 1 from 7).

1.   17 – 1 (from **7**,903 – **1**,997) = 16.
2.   9 – 9 (from 7,**9**03 – 1,**9**97) = 0.
3.   0 – 9 (from 7,9**0**3 – 1,9**9**7) =

In order to complete step 3, you must borrow and thus reduce step 2's answer by 1. Since that answer is 0, you can't reduce it. This means you need to borrow from step 1. This is the point of this exercise: **Subtraction Groups** can borrow from other **Subtraction Groups** in a way that causes a chain reaction of borrowing all the way to the left side of the problem.

The adjusted answer digits are now:

1. 17 − 1 (from 7,903 − 1,997) = **15**.
2. 9 − 9 (from 7,903 − 1,997) = 9.
3. 0 − 9 (from 7,903 − 1,997) = 1.

With this mess out of the way, you can find the difference of the rightmost **Subtraction Group**.

1. 17 − 1 (from 7,903 − 1,997) = 15.
2. 9 − 9 (from 7,903 − 1,997) = 9.
3. 0 − 9 (from 7,903 − 1,997) = 1.
4. 3 − 7 (from 7,903 − 1,997) = **6**.

This last calculation requires that you borrow from the third step. The adjusted answer digits are now:

1. 17 − 1 (from 7,903 − 1,997) = 15.
2. 9 − 9 (from 7,903 − 1,997) = 9.
3. 0 − 9 (from 7,903 − 1,997) = **0**.
4. 3 − 7 (from 7,903 − 1,997) = 6.

Your final answer is 15,906.

As you can see, subtraction involving one or more larger numbers is a skill that can require some time to master, although the basic principles are relatively easy to grasp. When watching someone with experience solving problems in this way, you'll likely notice their eyes darting about, switching focus between the **Subtraction Group** at hand and the neighboring **Subtraction Groups**. As you warm up to the idea of revealing answers from left to right, you'll become more comfortable with the cadence of the method.

Again, it must be stressed that if you're challenged not so much by the process itself but rather by the math involved in the individual **Addition/Subtraction Groups**, go back to focusing on being able to *immediately and instinctively* produce the sum or difference of any two numbers under twenty. You shouldn't move forward until you're lightning fast with that skill.

Compare these methods to classical addition or subtraction. The benefits are clear; any number you have to carry (**Overflow**, etc.) is going to have a value of 1—never more. This alone simplifies the method enough that—when the time comes to learn **Mental Calculation** and forego pencil and paper altogether—it won't seem so daunting to perform such calculations using only your mind.

Pre-Scanning

How would you handle the problem 28,941 − 18,000?

After looking at the numbers for a moment, you should realize it's a piece of cake. Every digit in the larger number is greater than or equal to its counterpart in the smaller number, so subtraction will be linear and painless.

How about 41,121 − 12,879? Before you even begin to perform this calculation, you can probably see why it's a total nightmare; almost every digit in the smaller number is larger than its counterpart in the larger number, so there will surely be a good deal of borrowing involved.

With time, you'll be able to look at a problem and immediately assess its level of difficulty and locate potential "problem areas" (**Overflow** or the need to borrow). This principle is called **Pre-Scanning**, and it can be applied to both addition and subtraction problems.

Integrate a quick **Pre-Scan** into your standard addition and subtraction processes; even though this technically increases the time needed to begin problems, it can improve your calculation accuracy and often saves time overall. For instance, if you **Pre-Scan** a problem and find no risk of **Overflow** or borrowing, you can tear through the problem from left to right quickly and with confidence.

Let's practice. **Pre-Scan** the following problems; quickly survey the two numbers involved and focus upon identifying the "problem areas." Don't worry about solving them.

1. $784,291 - 147,181$
2. $32,009 - 29,900$
3. $81,462 + 91,912$
4. $7,219 - 3,999$
5. $928 + 238$

Next, we'll talk about fundamental multiplication and division.

Quick Review

1. Adding/subtracting two larger numbers (five or more digits) can easily be done from left to right; however, take care to ensure that each pair of digits (called an **Addition** or **Subtraction Group**) is not at risk of being affected by neighboring pairs.

Lesson 4: Multiplication and Division

> "*I DISLIKE MATH, YET I RESPECT AND APPRECIATE THE FACT THAT MATH IS THE LANGUAGE OF THE UNIVERSE.*"
>
> —*LUCAS GRABEEL, SINGER/ACTOR*

By this point in life, we should all know what multiplication and division are and how they're used in daily activities. You'll do some intense work with these tools shortly. Before you get there, you have to do two things: first, make sure you're on very close terms with your basic multiplication tables. Then, become familiar with a basic concept called **Splitting**.

The Multiplication Tables

7 × 8! Go! Quickly!

What is the correct answer? More importantly, how long did it take you to get there? Just like you practiced until you could instantly recite the sum and difference of any two numbers under twenty, it's important to learn your multiplication tables (1 through 12) far better than you ever did in your youth.

If you don't remember your multiplication tables from 1 to 12, go learn them. Yes, now! Even the annoying parts—like 7 × 8 and 9 × 7—need to be memorized as thoroughly and instinctively as 2 × 2.

If you aren't lightning fast with your multiplication tables right now, that's perfectly understandable; while the more practical parts of the tables are used often, it's less common to touch upon the more obscure corners. Regardless, practice them before moving on.

×	2	3	4	5	6	7	8	9	10	11	12
2	4	6	8	10	12	14	16	18	20	22	24
3	6	9	12	15	18	21	24	27	30	33	36
4	8	12	16	20	24	28	32	36	40	44	48
5	10	15	20	25	30	35	40	45	50	55	60
6	12	18	24	30	36	42	48	54	60	66	72
7	14	21	28	35	42	49	56	63	70	77	84
8	16	24	32	40	48	56	64	72	80	88	96
9	18	27	36	45	54	63	72	81	90	99	108
10	20	30	40	50	60	70	80	90	100	110	120
11	22	33	44	55	66	77	88	99	110	121	132
12	24	36	48	60	72	84	96	108	120	132	144

Continue on only when you feel you've mastered the above multiplication tables.

Splitting

Now let's talk about **Splitting.** The concept of **Splitting** is integral to many of the multiplication and division methods you'll soon learn.

To **Split** a number is to divide it by two. Yes, that's all; there's no catch. This is easy for some numbers; 66 **Split** is 33, 46 **Split** is 23, and 12 **Split** is 6. Even some odd numbers, such as 41, **Split** easily and without much thought required (41 **Split** equals 20.5). However, numbers (especially odd ones) become more challenging to **Split** as they grow larger.

How would you split 93? The trick is to break the number into manageable pieces (93 can be broken down to 90 and 3) and work from left to right; half of 90 is 45, and half of 3 is 1.5. Take the sum of the two partial numbers: 45 + 1.5 = 46.5.

Verbose Splitting

93 was easy, you say? Okay – how about 3,917?

Don't panic or start guessing. You'll **Split** this in the same way, except this time, you'll piece your partial answers together as you come to them. Half of 3,000 is 1,500. Half of 900 is 450. Piece these two partial answers together: 1,500 + 450 = 1,950. That's the answer so far. Moving on, half of 17 is 8.5. Now, piece *these* two partial answers together: 1,950 + 8.5 = 1,958.5. Given your ability to immediately produce the sum or difference of any two numbers under 20, the addition here should be a breeze.

This is called **Verbose Splitting** because it involves breaking the number down and acting upon each part in turn.

Let's split 4,381.

Half of 4,000 (2,000) plus half of 300 (150) equals 2,150. Half of 80 (40) plus 2,150 is 2,190. Half of 1 (.5) plus 2,190 is 2,190.5.

Split the following numbers using **Verbose Splitting**.

1. 7,626
2. 1,387
3. 4,909

Answers:

1. 3,813
2. 693.5
3. 2,454.5

Short Splitting

When a number comprises several easily-**Split** digits (or pairs of digits) sitting side-by-side, you may choose to break the number up and **Split** them individually. For example, to **Split** 1,662 using **Verbose Splitting**, you would begin by **Splitting** 1000, followed by 600, then 60, and then 2 (four steps). However, this number is composed of two numbers, sitting next to each other, that are easily split: **16**62 and 16**62**. Thus, you may begin by **Splitting** 16, then **Split** 62, resulting in 8 and 31, or 831 (two steps).

Always do this when each **Split** chunk results in a number that will not need to interact with its neighboring chunks. For instance, while 1,662 breaks into two numbers that **Split** evenly and are self-contained (16 becomes 8 and 62 becomes 31), 1,560 does not, as 15 **Split** is 7.5. The .5 would need to be worked into the neighboring chunk

(60 ÷ 2). Because of that, **Short Splitting** is off limits in this case. Solve this in whatever way is comfortable for you. This could mean full **Verbose Splitting** ("half of 1,000 is 500, half of 500 is 250, and 500 + 250 = 750. Half of 60 = 30, and 750 + 30 = 780") or taking a partial shortcut ("half of 1,500 is 750, half of 60 is 30, and 750 + 30 = 780").

Time to Practice

Splitting isn't so bad if you take your time and follow a process. No panicking or guessing; once you're comfortable with numbers, you can play with them creatively. **Split** the following numbers in your head as quickly as you can:

1. 39
2. 71
3. 892
4. 1,290
5. 2,534
6. 998
7. 6,619

Answers:

1. 19.5
2. 35.5
3. 446
4. 645
5. 1,267
6. 499
7. 3,309.5

Think about how you approached the first number. Barring any shortcuts, you should have said something like, "half of 30 is 15, and half of 9 is 4.5. 15 plus 4.5 equals 19.5."

How did you solve problem 3? Barring any shortcuts, you should have said something along the lines of, "Half of 800 is 400, half of 90 is 45. So far, you have 445. Half of 2 is 1. 445 plus 1 is 446."

How did you solve problem 5? Barring any shortcuts, you should have said "half of 2,000 is 1,000, half of 500 is 250, and half of 30 is 15, so you have 1,265. 1,265 plus half of 4, which is 2, is 1,267."

With practice, you'll be able to determine a number's **Split** value upon seeing it without spending as much time going through the steps. Using the same three examples:

For problem 1 (39), you'd simply say, "15...19.5."

For problem 3 (892), you'd say, "400...445...446."

For problem 5 (2,534), you'd say, "1,000...1,250...1,265...1,267."

Splitting will be needed for more advanced multiplication and division methods, so get used to it now. Begin splitting numbers that you see in your daily life. Speed comes with practice.

Doubling

As you can guess, to **Double** a number means to multiply it by two. This is simple for comfortable, common numbers (16 × 2 = 32) and those that **Double** easily. For some numbers, doubling the entire number simply means **Doubling** each individual digit or chunk. Take $133 \times 2 = 266$ for example: doubling each digit, you have $1 \times 2 = 2$, $3 \times 2 = 6$, and $3 \times 2 = 6$; and doubling it in two chunks, you have $13 \times 2 = 26$ and $3 \times 2 = 6$.

Doubling numbers that involve carrying makes things a bit more difficult. How would you double 168?

The best approach is to first **Double** the two leftmost digits (16). Store the answer (32) in your mind for a moment and look at the final (ones place) digit, 8. **Doubling** this will cause you to carry a 1, so first increment your 32 to 33, and then use only the ones place digit from the 8×2 calculation (the 6 from 1**6**). Your answer is 33 followed by a 6, or 336.

This may seem like an obvious exercise for many of us, but the principles are not intuitive for everyone, and they're important to understand on a fundamental level. You'll have to **Double** some daunting numbers in the coming sections.

Double the following numbers in your head as quickly as you can:

1. 69
2. 102
3. 512
4. 898
5. 2,298
6. 23,944

Answers:

1. 138
2. 204
3. 1,024
4. 1,796
5. 4,596
6. 47,888

Practice Before Moving On

It's vitally important that you have a firm grasp on each lesson in *Master the Language of the Universe* before moving forward. Many later skills rely and build upon these fundamental ones. Feel free to spend a few extra days working on your multiplication tables. Get in the habit of quickly **Splitting** and **Doubling** numbers you encounter in your daily life; **Split** and **Double** the current time, license plates, phone numbers, addresses, grocery store totals, and the like.

Quick Review

1. Mathematical fluency requires expert-level familiarity with every possible two-number multiplication problem involving only zero through twelve (the multiplication tables). You must be able to produce these answers in a confident, instant, and reactionary way.
2. To **Split** a number (cut it in half), first break it up into manageable components. **Split** the components one-by-one from left to right, and do so calmly (resist the urge to guess or panic). **Splitting** is a vital aspect of some more advanced methods you'll learn.

3. To **Double** a number (multiply it by 2), first break it up into manageable components. **Double** the components one-by-one from left to right, and do so calmly (resist the urge to guess or panic). **Doubling** is a vital aspect of some more advanced methods you'll learn.

Lesson 5: Roughing It (the Art of Quick, Imprecise Calculations)

"OMG HELP STUPID MATH QUESTION?...YOU WILL BE A HERO IF YOU ANSWER THIS WITH AND EXPLANATION SO PLEEEAAASE HELP ME"

—TITLE OF A POST ON YAHOO! ANSWERS (CANADA)

This skill will most likely prove to be the most useful one you'll learn in this entire book.

Most of the time, you don't need to know that 40 goes into 1,236 exactly 30.9 times; you just need to know that if you have 40 classrooms, 1,236 students, and a maximum capacity of 20 students per room, you've got a problem. This is when quick, imprecise calculations come in handy. Let's learn some tricks for squeezing as much precision as you can out of loose calculations.

Ignore and Compensate

To use this trick, ignore the numbers' smaller digits (usually the ones places or both the tens and ones places), and then make up for this by adjusting the answer based on the operation and direction you rounded.

Whenever you round down for addition or subtraction problems, you'll know to round your final answer up. If you round up in order to solve the problem, you'll know to round your final answer down. Take 4,809 + 664 for example. You can make two determinations with certainty:

The answer is "roughly" 5400: You know that 48 + 6 (from **4,8**09 + **6**64) is 54, so use this as the base of your rough answer.

The actual answer is going to be higher than the rough answer: Since you rounded 4,809 to 4,800 and 664 to 600 (both down), you must round final answer up, so you can say the final answer is "a bit more than 5,400." You'd be right: the precise answer is 5,473.

Now that you've done an addition example, let's look at a simple subtraction example. 554 − 441 equals 113. If you subtracted 44 from 55, you'd get 11 as the first two digits of your initial rough answer. Accounting for the fact that you rounded *down*, you should adjust the final answer *up*, to "a little more than 110," which is accurate.

This method also applies to multiplication problems; here, too, the final answer is adjusted in the direction opposite the rounding. How would you handle 61 × 91? If the circumstances allow, you can rough it rather than perform the actual calculation; 6 × 9 = 54, so the answer could be expressed as "a bit more than 5,400." It's important to note that when dealing with multiplication or division, the numbers lost in rounding can be significant. If the situation calls for a greater degree of precision, you must account for this.

Let's look at 61 × 91 again, but this time more carefully. Given that you rounded down on both counts (to 60 × 90), you can assume that the actual answer is a fair amount higher than 5,400. In fact, using 91 instead of 90 would alone bring the answer up to almost 5,500 without even taking into account the 1 we ignored when rounding 61 down to 60. If you took the time to figure this out, you could instead say that the answer is "more than 5,500" (the real answer is 5,551). This illustrates an important point: *The work you need to put in really depends on the precision required by the situation.*

Division is more complicated, because minute changes in either number involved can cause major fluctuations in the answer. Let's go back to the classroom/students problem presented in the introduction to this lesson: 1,236 ÷ 40.

Since 4 goes into 12 (from **4**0 into **1,2**36) exactly 3 times, the rough answer will begin with a 3. The smaller number (40) is two digits shorter than the larger number (1,236), so the answer will be either be two or three digits long. To figure out which, simply ask if the dividend (12) is more or less than ten times the divisor (4). Since it's less, the answer will in most cases be two digits long. From this information, you can conclude that the answer is "roughly" 30. The actual answer is 30.9, so this seems like a reasonable response.

Another division example: how should you handle 850 ÷ 123? The actual answer is 6.91, but let's look at the answers you get when addressing this problem with different precision tolerances:

- 8 ÷ 1 = 8
- 85 ÷ 12 = 7

With the actual answer being 6.91, would "about 8" have been acceptable? How about "around 7?" Again, it depends heavily upon context of the problem. This example illustrates another important point: *When it comes to multiplication and division, this method is best used for problems involving smaller numbers.*

Using the **Ignore and Compensate** method, do the following computations in your head. The correct answers are listed below. Don't cheat yourself; figure out and write down the answers, then check them. Once finished, if there were any mistakes, try to figure out where you went wrong. Try to get as close to the correct answer as possible without spending too much time or brainpower.

1. 65×44
2. 82×15
3. 190×16
4. $440 \div 23$
5. $891 \div 66$
6. $333 + 741$
7. $186 + 230$
8. $8{,}392 - 901$
9. $764 - 88$

Answers:

1. 2,860
2. 1,230
3. 3,040
4. 19.13
5. 13.5
6. 774

7. 416
8. 7,491
9. 676

One Up/One Down

Let's explore a quick and easy method for addition. How would you quickly handle $6,431 + $1,777? You could use the method taught above, but since one number rests near the "bottom" of a "thousand span" (meaning 6,431 is closer to 6,000 than it is to 6,999) and the other number rests near the "top" (1,777 is closer to 1,999 than it is to 1,000), a rough estimate could be made by rounding one number down and the other up, beginning with the same digit or "place" in each (in this case, the hundreds place). Either direction works; you could round this to $6,400 + $1,800 or $6,500 + $1,700, both of which leave you with an answer of $8,200. An answer of "around $8,200" may be close enough to the actual answer, $8,208, depending on the context of the problem.

This method is obviously more precise if the two numbers are comparably distant from their rounded versions; for example, in the problem 331 + 269, both numbers are exactly 31 away from their nearest rounded counterparts. In some cases, however, one of the numbers may be close to the bottom of the hundreds place (e.g., 402) and the other closer to the middle (e.g., 458); no matter—this method is still great for extracting superficial precision from a quick calculation.

This won't work if both numbers reside near the top (1,392 + 296) or bottom (1,329 + 226) of a number span (in this case, the hundreds place). For this, round both numbers either up or down, and then compensate as you learned before.

Using the **One Up/One Down** method whenever possible, do the following computations in your head. The correct answers are listed below. Don't cheat yourself; figure out the answers, write them down, and then check them. Once finished, if your rough answer is unreasonably distant from the actual answer, try to figure out where you went wrong. Try to get as close to the correct answer as possible without spending too much time or brainpower.

1. 23 + 438
2. 75 + 213
3. 754 + 7,892
4. 8,243 + 907

Answers:

1. 461
2. 288 (candidate for **One Up/One Down**)
3. 8,646
4. 9,150

Multiplication Between Two Targets

When a *very* rough answer will suffice, this multiplication method gets you an answer quickly.

Let's learn by walking through an example: 65 × 271. You could round one of the numbers, so let's say you choose to round 271. This number lies between 200 and 300, so you have two potential ways to round. Your answer will lie somewhere in between 65 × 200 and 65 × 300. Ignoring the zeroes for a moment, multiplying 65 × 2 gives you

130 and 65 × 3 gives you 195. Since 271 is closer to 300 than 200, place your rough guess somewhere slightly closer to 195 than 130. Let's say 170. Then add the two zeros back to the end of your answer. This turns the answer into "around 17,000."

How close did you get? The real answer is 17,615.

Let's try another: 124 × 28. First round 124 up to 125 to make it easier to work with. As 28 is closer to 30 than 20, multiply 125 × 2 (250) and 125 × 3 (375), and choose an answer that is much closer to the latter (say, 340). Adding the stripped 0 back in, you end up with "around 3,400." The real answer is 3,472.

It's important to use two targets because it gives you a sense of the scale of adjustment needed. Had you simply multiplied 125 × 30 and knew that you had to "drop down a bit," you would have no sense of *how much* you needed to drop. But using two targets, you quickly realize that 28 is 2/10 of the way from 30 down to 20. With this knowledge, you can "come down roughly 2/10 of the way between 375 and 250," which, though imprecise, is something you can grasp with practice.

We rounded 124 to 125 because it was easier to work with (due to currency, you should be used to dealing with numbers that end in 25). This illustrates an important point; once you're comfortable with numbers, finding rough answers is more art than science. Get creative; find your way to a rough answer in any way that makes sense to you. You can simplify numbers as much as you'd like before performing calculations, as long as you adjust afterward.

If you were multiplying 81 × 5,300, you could choose to multiply 8 × 5 and 8 × 6 (answers: 40 and 48, respectively), and add the zeroes back in (400,000 and 480,000). Now you know the answer is "somewhere in the 400,000's but less than 480,000." Want more precision? Simply adjust for the fact that the 5,300 is a bit closer to 5,000 than it is 6,000, so the answer should be closer to 400,000 than 480,000. What do you think, maybe around 430,000? With the actual answer being 429,300, you'd be in good shape.

Using the **Multiplication Between Two Targets** method, do the following computations in your head. You'll find the correct answers below. Don't cheat yourself; figure out and write down the answers, then check them. Once finished, if your rough answer is unreasonably distant from the actual answer, try to figure out where you went wrong. Try to get as close to the correct answer as possible without spending too much time or brainpower.

1. 32×12
2. 83×28
3. 28×29
4. 17×55
5. 65×891

Answers:

1. 384
2. 2324
3. 812
4. 935
5. 55242

Quick Review

1. When making rough, imprecise calculations, several strategies
    exist that can easily improve the precision of the answer,
    including the **Ignore and Compensate, One Up/One Down**,
    and **Multiplication Between Two Targets** methods.

Review and Development: Section 1

In *Section 1*, your goal was to develop and reinforce simple math skills that you will build upon shortly. Relative to what you'll learn next, this section has been low-impact, so take advantage of this time to ensure you're comfortable with all the lessons and mentally prepared for what lies ahead.

Review

Let's take a look at what you learned in *Section 1*.

1. **Mathematical Fluency**: The first step toward achieving mathematical fluency is to become incredibly familiar with every possible addition and subtraction relationship involving numbers under twenty. The ability to produce these sums and differences in a confident, instant, and instinctual manner better prepares you to perform more difficult addition and subtraction problems quickly.

2. **Familiar Numbers**: Using **Familiar Numbers** (rounding numbers up or down to their nearest multiple of ten), you can tackle daunting addition or subtraction problems with ease.

3. **Single-Offset Addition**: **Single-Offset Addition** involves rounding one of a problem's two numbers (usually the smaller number) and accounting for the **Offset**, or the amount by which it was rounded.

5. **Dual-Offset Addition**: **Dual-Offset Addition** involves rounding both of a problem's numbers and accounting for their **Offsets**, or the amounts by which they were rounded.

4. **Addition Subtraction**: **Addition Subtraction** is a method of subtraction by which you split a difficult subtraction problem into two smaller problems: a simpler subtraction problem and an addition problem.

5. **Dual-Offset Subtraction**: **Dual-Offset Subtraction** is a method of subtraction by which you round both numbers in the same direction, and then subtract the **Final Offset** from the pre-**Offset** answer.

6. **Left-to-Right Addition/Subtraction**: Adding/subtracting two larger numbers (five or more digits) can easily be done from left to right; however, take care to ensure that each pair of digits (called an **Addition** or **Subtraction Group**) is not at risk of being affected by neighboring pairs.

7. **Multiplication Tables**: Mathematical fluency requires expert-level familiarity with every possible two-number multiplication problem involving only zero through twelve (the multiplication tables). You must be able to produce these answers in a confident, instant, and reactionary way.

8. **Splitting**: To **Split** a number (cut it in half), first break it up into manageable components. **Split** the components one-by-one from left to right, and do so calmly (resist the urge to guess or panic). **Splitting** is a vital aspect of some more advanced methods you'll learn.

9. **Doubling**: To **Double** a number (multiply it by 2), first break it up into manageable components. **Double** the components one-by-one from left to right, and do so calmly (resist the urge to guess or panic). **Doubling** is a vital aspect of some more advanced methods you'll learn.

10. **Rough Calculations**: When making rough, imprecise calculations, several strategies exist that can easily improve the precision of the answer, including the **Ignore and**

Compensate, **One Up/One Down**, and **Multiplication Between Two Targets** methods.

Development

Spend seven days practicing the skills covered in *Section 1*.

Days 1 – 2

Spend the first two days solving addition problems using **Single-** or **Dual-Offset Addition.** Choose a method based on your preference and personal comfort level with each. Please keep in mind that you will be using the chosen method regularly, and thus advancing and becoming even more comfortable with it; take care to choose the method that will serve you best in the long run, even if it may be challenging right now. Using a pencil and paper, spend time each day coming up with two arbitrary numbers and then adding them together. Record the problems and answers, but never your work.

Day 3 – 4

Spend the next two days solving subtraction problems using **Addition Subtraction** or **Dual-Offset Subtraction.** As with addition, choose the method that makes the most sense to you, and then spend a few hours devising and solving problems using a pencil and paper. Again, record the problems and answers, but never your work.

Day 5

Day 5 is dedicated solely to the further reinforcement of the multiplication tables for the numbers one through twelve. You should already have this down, otherwise you shouldn't have moved on, but there's always room for improvement. Find any remaining trouble areas and work to reinforce confidence through focused repetition.

Day 6

Spend *Day 4* **Splitting** and **Doubling** every number you see. You did this when you first touched upon these subjects, but now you're going to spend a day focused entirely on this. Use license plates, parts of phone numbers, addresses, bus or train numbers, the time, the four-digit groupings of credit card numbers, etc. Numbers are all around, and if it's a number, you can **Split** and **Double** it.

Additionally, spend dedicated time today—three half-hour blocks, if possible—using a pencil and paper to scribble down arbitrary two- to five-digit numbers to **Split** and **Double**.

Day 7

Using the three methods discussed (**Ignore and Compensate**, **One Up/One Down**, and **Multiplication Between Two Targets**), scribble down quick, arbitrary math problems—addition, subtraction, multiplication or division—and solve them as quickly as possible using the best method for each. Throughout the day, you should find yourself becoming more familiar with the methods, solving problems using creative approaches, and finding a comfortable personal middle-ground between speed and precision.

Before wrapping up, spend extra time on any part of this section you aren't totally comfortable with. Once you're ready to continue, you're going to learn how to handle some common mathematical scenarios you may find yourself facing.

Section 2: Practical Mathematical Skills

Well done! You can now add, subtract, multiply, and divide in ways that render you fully equipped to handle most day-to-day practical math problems. In *Section 2*, we'll discuss some core mathematical concepts that are both commonly misunderstood and widely applicable to daily life. This includes:

1. **Guestimation**: **Guestimation** is the art of making estimates based on incomplete information. Human beings are notoriously bad at this, but there are some quick and easy ways to increase accuracy and precision.

2. **Probability**: When it comes to life applicability, **Probability** is one of the most important mathematical disciplines out there; unfortunately, it's also one of the most universally misinterpreted. We'll address a few common misconceptions and learn how to apply **Probability** principles to your life.

3. **Options**: Life consists of a series of forking possibilities; here you'll learn the basic math behind this phenomenon.

4. **Percentages**: You deal with percentages on a daily basis. The ability to grasp the concept of "how much of a whole" is vital to understanding the world—we operate within a conceptual model that's built upon individual objects or ideas that manifest themselves as either wholes or parts of wholes.

5. **Value Comparison**: **Value Comparison** is an almost constantly-used skill, and yet most people are generally

untrained in its theory and unskilled in its application. You'll learn how to take two values (even if they are measured differently) and compare them efficiently.

6. **<u>Sorting</u>**: Many aspects of your life need to be organized. By understanding some simple sorting principles and methods, you can become much better at this.

Let's begin.

Lesson 1: Guestimation

Long before Sherlock Holmes was ever portrayed on film, Sir Arthur Conan Doyle wrote for him a similar (and arguably more poetic) line: "I never guess; it's a shocking habit, destructive to the logical faculty." You'll soon learn exactly how destructive a guessing habit can be, as well as how easy it is to overcome.

In the last section, you learned several methods for coming up with quick, rough answers to common types of math problems. Something these methods have in common: they are less "wild guess"-focused and more "collect and use as much data as possible in the shortest amount of time"-focused. Ultimately, the best guess is one that minimizes the need to guess. Extending this concept, let's get away from math problems and learn to **Guestimate** quantity.

Guestimation—an adorable word spawned from a love affair between "guessing" and "estimation"—is a skill that's needed fairly often, but is seldom executed skillfully. While guessing implies a blind, uncalculated shot in the dark, estimation refers to a more scientific approach; **Guestimation** is a hybrid technique—the type of approach you're afforded when faced with limited data, time, or both.

Humans' inability to **Guestimate** well manifests itself as a general failure to predict the probability of future events, estimate quantity, and interpret the passing of time. Though natural instincts are usually misleading when it comes to **Guestimation**, precision can be drastically increased by following a few simple pseudo-scientific pointers.

Forgoing discussion of time or event likelihood, let's focus on **Guestimating** pure quantity (groups of objects).

Learning to properly **Guestimate** is useful for literal, practical **Guestimation** tasks, but it also helps ingrain positive habits; the practice of collecting data (no matter how rough) and learning to distrust your gut reactions will prove useful in many areas of life. Moreover, **Guestimating** with restraint will increase others' impression of your overall competency; your restraint will come across as evidence of a rare, analytical, non-impulsive disposition.

How to Guestimate Quantity

One of the simplest applications of **Guestimation** is to determine the number of uniform, static objects occupying a two dimensional plane.

Small Groups of Uniform, Static Objects on a Two-Dimensional Plane

In this context, "objects" refers loosely to any number of physical, viewable things, such as letters in a word, people in a group, stars on a flag, or buttons on an ATM. "Uniform" means "roughly the same" (groups of objects of consisting of relatively similar sizes or types), and "static" means "unchanging." Though the term "two-dimensional plane" is used here, this doesn't necessarily only apply to flat surfaces; rather, it applies to any situation in which all involved objects/entities can be seen from a single perspective. A good rule of thumb is that if you must view the containing setting from a secondary angle to see all the objects, it is not applicable to this context.

These criteria could therefore describe written text, the number of pages in a book (viewed from the side), the spokes on a hubcap, the crayons in a pack (viewed from above), and so on. This would *not* include something like a jar of gumballs, as all of the objects (gumballs) cannot be seen from a single vantage point (one side of the jar).

First you'll learn to **Guestimate** small groups of such objects, then work your way up to larger groups. The principles involved in assessing both small and large groups of said objects are very similar.

Human beings come pre-loaded with an intuitive ability to instantly recognize the number of objects in a small group (five or fewer objects) without counting them. This is called **Subitization**, and—unless you suffer from a specific condition called *dyscalculia*, which impedes this ability—it probably comes naturally to you. Think about it: when someone holds up three fingers, do you count "one...two...three" fingers, or do you just see "three" fingers? Do you ever have to count the number of fighters in a boxing ring, or the number of letters in the word "car?" **Subitization** has obvious evolutionary/survival functions ("there are four timber wolves attacking us, and I am one of only two remaining tribesmen") and requires no conscious effort.

Any group with more than five objects requires you to use your mind a bit, and the more objects there are, the more you can benefit from **Guestimation**.

There are a few things to you should do when **Guestimating** the number of objects in a group. Most importantly:

Gather (At Least *Some*) Data: The trick to quality **Guestimating** is to rely as little as possible on your instincts, and gather as much actual data as you can as quickly as possible. A "rough count" is better than "no count," and a "thorough count" is better than a "rough count," so do whatever you can in the time you have. Don't let your ego overpower your logical faculties. The goal isn't to improve your gut feeling, but to quickly and automatically make a rough *educated* guess.

I asked eleven people to take a look at the paragraph you just read (beginning with "The trick to...") and guess the number of words it contained within six seconds. Some people overshot the actual number by a significant margin, with a few of them guessing well over one hundred and twenty words. Others drastically underestimated. Very few were even close to the actual number. However, by asking the same people to count the number of words on a single line and then multiply this rough average by the number of lines, all eleven participants did much better. Half of them **Guestimated** within ten words of the correct answer (82 words).

Don't Panic: When **Guestimating**, there is often very little time to gather data, and so you may be tempted to dismiss the first piece of advice and rely on instinct. This is a bad idea. Remain as calm and composed as possible.

Scrutinize Your Mistakes: Imagine passing a road sign or seeing an unfamiliar name, and attempting to **Guestimate** the number of letters it contains. You'd probably be off by a few (after all, you're new at this), but an important way to improve is to determine the correct answer afterward and scrutinize your error. Try to figure out *why* your guess was off from the correct answer. Did you fail to collect adequate information? Did you rely too much on your instincts? Were the characters unusually close together or far apart (a phenomenon called kerning or mortising, which can sometimes create the illusion that there are more or fewer letters than there actually are)? Were you thrown off by the words spanning several lines?

Try to **Guestimate** the number of pages in a nearby book (while looking at the book from the side). After finalizing your answer, find the correct answer. Were you over or under? Why? Are the pages thicker or thinner than expected? Are you simply inexperienced at relating book thickness to page count?

Once you've come to a conclusion about the reasons for your inaccuracy, try to embed this lesson deep in your mind, so you'll be more likely to consider it next time. Slowly, your ability to guess the number of objects in groups (whether letters, words, pages, or anything else) will improve.

Practice honing this skill on anything you can **Guestimate** quickly and then check afterward. You can use a small moving group of people, floral lines on a theatre curtain, words on a page, or cars in a parking lot. Begin with the number of keys on your keyboard or the number of pages in another book. How about the number of words in this paragraph?

Spend some time practicing, as it's important to hone this skill a bit before tackling larger groups of objects, which we'll discuss next. Feel free to stop here and spend up to a few days working on this skill before moving forward.

Extending Guestimation to Larger Groups of Objects on a Two-Dimensional Plane

Using these strategies (collecting data, resisting panic, and scrutinizing mistakes), let's move on to larger groups of objects on a two-dimensional plane. As with the "words in the paragraph" example, larger groups of objects require you to take a small sample (a part of the whole), determine how many objects exist within this sample, and then perform rough calculations upon this sample to extrapolate it to an approximation of the whole. The difference between smaller and larger groups is simply that larger ones sometimes necessitate different or additional operations.

Imagine an eagle's-eye-view photograph of a concert. You want to estimate how many people there are in the crowd. First, try to quickly assess how many rows of people stand between the stage and the rear wall of the concert hall. People don't stand neatly aligned with each other, and thus this count varies depending on which area you selected as the sample set. Try to select an "average" row. Let's say you count 51 people.

Next, perform a second count, this time taking a sample count of the columns—the people between the left and right walls. This time, let's say you count 40. So, 51 (from front to back) × 40 (from side to side) = 2,040 concertgoers.

Before settling on an answer, account for any oddities or anomalies. Imagine that there are two patches within the crowd that are less densely packed. One of these areas is slightly thin, and the other is very thin. To account for this, you might round down from 2,040 to about 1,900 (a difference of 140 people).

Assuming you have limited time and resources, this method would produce as good a **Guestimate** as you could hope for, allowing you to come within a reasonable margin of the true total. Though far from perfect, the results are superior to a simple guess. With a little more time, you can take multiple sample row counts and average them (three sample rows containing 51, 54, and 47 people averages out to about 51). You could then do the same for the columns to further increase your accuracy.

In this example, you extracted samples of the two dimensions, multiplied them, and accounted for anomalies in order to come up with a **Guestimate**.

If you were to **Guestimate** the number of characters in this paragraph, you could count the number of characters on a single line and multiply it by the number of lines. However—assuming there's some wiggle room in terms of precision—it would be faster to count the number of words in a line, multiply it by the number of lines, and multiply the outcome by a loose average of the number of characters per word. Though words vary greatly in length, this will decrease the time needed to come up with a useful **Guestimate**.

In the above paragraph, the actual number of words is 95 and the number of characters is 425, meaning the average word contains 4.47 characters. When asked to perform this calculation using the method described above, the same eleven participants from the first test ascertained that there were about 88 words in the paragraph at an average of 5 characters per word, which equals 440 characters.

440 (**Guestimate**) versus 425 (actual count). So, quickly **Guestimating** resulted in 96.5% of the precision of counting each individual character.

*Tip: **Guestimation** of larger groups of objects, like smaller groups, is a skill that is improved by strict self-scrutiny. The harder you are on yourself when practicing, the more likely you will be to remember and reference your mistakes in the future.*

Objects Occupying a Three-Dimensional Space

We've all seen the "jar of gumballs" challenge. It could be jellybeans, peanuts, or marbles, and you may have seen it at a carnival, supermarket, or state fair. Regardless, the concept is the same; guess the number of objects, and if your guess is the closest to the real number, you win a teddy bear/car/lifetime supply of diapers.

It's human nature to want that prize, so let me guess—you've attempted this challenge and failed. I'd also guess that you were surprised by the actual answer. You can get better at this type of estimation, and in doing so, become better equipped to face its more practical applications.

The gumball example—a three-dimensional version of the "concert crowd" problem—is essentially a question of volume (remember geometry? Sophomore year? No? Nothing?). The unit of volume in this problem is the gumball, or more specifically, the three-dimensional space taken up by one gumball, including the air occupying the crevices surrounding it.

Using the image shown here, first figure out roughly how many "gumballs wide" the jar is at its thickest point. Though the gumballs are sitting on top of each other somewhat haphazardly, it looks to be about six (five to seven). This is the diameter.

The formula for volume (of a cylinder) is pi × radius² × height. If the diameter is six, then the radius (half of the diameter) is three.

Next, count how many "gumballs high" the jar is at its tallest point. It appears to be about thirteen. However, there is one factor to consider that calls for an adjustment of this number; the jar narrows at the top, beginning about 1/6 of the way down. It may be prudent to lower your loose height **Guestimate** to about ten.

Following the formula of pi × radius^2 × height, the volume should be about 283 (3.14 × [3^2] × 10).

The actual number of gumballs in the pictured jar is 297, so this **Guestimate** is very much in the ballpark!

Though this illustrates the idea of giving some thought to the problem and doing some rough math instead of purely guessing, this method is way too complicated. Let's simplify it.

Though the container above is cylindrical, imagine that it is instead a rectangular prism (pictured to the right) with flat sides and corners. The formula to calculate volume of a rectangular prism is considerably simpler: side$_1$ × side$_2$ × side$_3$. This eliminates the need to deal with pi or exponents.

In this case, 6 (width/diameter) × 10 (height) × 6 (depth, which will be the same as the width) = 360.

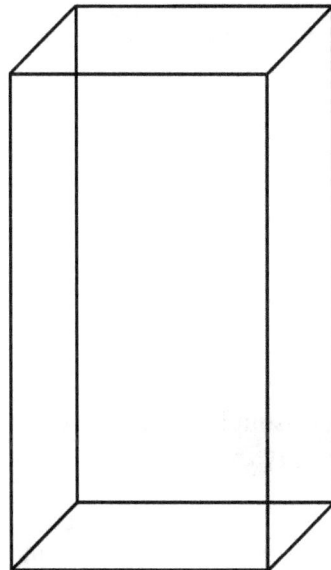

A cylindrical space holds less volume than a rectangular prism with the same height and width, so any estimate of the actual number of gumballs in such a jar must also decrease.

The images to the left show two-dimensional cross-sections of both shapes (ignoring height) at different scales; the top has a 5" diameter, the bottom 100". Think of three-dimensional spaces—like a space where you'd store gumballs—as "stacks" of these two-dimensional slices. In both cross-sections, there's roughly a 20% difference between the area of the square version and the area of the circular version. The square is always the larger of the two.

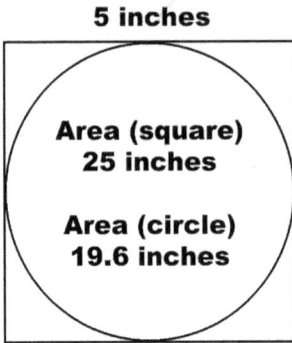

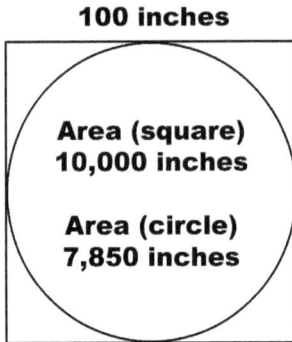

5 inches

Area (square)
25 inches

Area (circle)
19.6 inches

100 inches

Area (square)
10,000 inches

Area (circle)
7,850 inches

So, when **Guestimating** the volume of a cylindrical container by treating it like a rectangular prism in order to simplify your math, decrease your **Guestimated** answer by about 20%. This turns 360 into about 288. Using the complex cylindrical formula, your **Guestimate** was 283. Therefore, the much simpler "(side$_1$ × side$_2$ × side$_3$) minus twenty percent" method altered your answer by only about 1%.

Since this method simplifies the calculation, consider using it when **Guestimating** the volume of a cylindrical three-dimensional space. Note that if you were asked to **Guestimate** the number of gumballs in a rectangular prism (a container with hard edges and 90-degree angles), you would of course perform the $side_1 \times side_2 \times side_3$ calculation without detracting the 20%.

The Point: Accept Your Limits and Collect Data

When people see a jar similar to this and are asked to guess how many gumballs are inside, guesses can be all over the map—as low as 50 and as high as 750. Regardless of how absurdly high or low someone's data-free instinctual estimate may be, human arrogance prevails; the guesser usually announces his or her answer with confidence, and is dumbfounded that the arbitrary, reactionary guess was inaccurate.

The above examples teach you how to address specific situations, and what types of calculations to perform for each. Keep in mind that the appropriate methods for collecting quick, rough measurements are different for every type of problem you may encounter. *The point isn't just to learn how to guess the number of gumballs in a jar, but to instead show you how to approach problems of any kind.* I want you to take away from this lesson the understanding that your guess—your intuition, your heart's answer—is garbage. Get over it.

Collect data and use whatever science you know as quickly as you can.

Quick Review

1. When performing any sort of estimation, it's important to resist the urge to guess or rely on intuition. Force yourself to take measurements and use any available data to make rough calculations. This partially-informed type of guessing is called **Guestimation**.

Lesson 2: Practical Probability

Many years ago, I was an occasional participant in a standing weekly poker game. Through this game, I came to know a friend-of-a-friend who could look at the outcome of a hand of cards and say something like, "Wow, tough luck. The chances of that happening are roughly one in eighteen." For all I knew, he could have been lying; in those days, my grasp on **Probability** was so poor that I wouldn't have known. However, when prompted, he could explain the calculations behind his conclusions. His explanations taught me that the math he was using wasn't difficult; it only seemed so because most people don't have a clue how to approach **Probability**.

Probability is applicable to almost every aspect of daily life but remains widely misunderstood. It's barely covered in high school and has the reputation of being an intimidating and boring branch of mathematics. This reputation isn't entirely undeserved, as **Probability** is a complex, multi-dimensional discipline that can take years to fully understand; however, you can understand the core concepts and dispose of some common misconceptions in minutes, as the parts most commonly applicable to daily life are actually fairly simple.

What is Probability, and Why Do I Need to Understand It?

Mathematicians such as Pascal and Fermat first explored modern **Probability** many years ago, and it soon blossomed into a respected branch of mathematics, with applications ranging from computer science and risk assessment to gambling and office water cooler sports chat. **Probability** allows you to understand the likelihood that an event will occur, assuming you don't know for certain whether or not the event will *definitely* occur.

It's no wonder that most people—having only a superficial understanding of **Probability**—tend to overlook its value when assessing a situation. Most people have a tendency to disregard math and rely on gut instincts, but familiarity and comfort with basic **Probability** provides a helpful perspective in day-to-day situations. You'll be able to better understand the odds of an event occurring based on certain factors, such as your knowledge of past occurrences. As with **Guestimation**, you need to put less stock in your intuition and instead collect as much data as possible. You can then use simple mathematical tools to calculate **Probability** and make more informed decisions.

If everyone truly understood **Probability**, casinos would be far less busy and Nostradamus' predictions would inspire much less awe. A basic understanding of **Probability** can put the scale of the world and your actions into perspective, which can sometimes help to put your mind at ease. People with limited knowledge of **Probability** greatly overestimate the odds that they are at risk of certain diseases, violent crime, and airline accidents (in fact, you are more likely to die at an intersection on the way to the airport than in a plane crash).

Unlike almost any other explanation of **Probability** you may see, we're going to approach this topic without learning a single formula. We'll also use some unique terminology that makes the topic less daunting, breaking otherwise-complex rule sets into simple "ors" and "ands."

To demonstrate these concepts, we'll return to the same two examples repeatedly. The first is the classic example of a die (as in, singular for "dice"). The second is an example in which you possess 100 tennis balls of varying colors.

Basic Probability

Let's quickly compute the likelihood of an occurrence. **Probability** is generally measured on a scale of 0 (definitely not happening) to 1 (definitely happening). Anything in between means, "it could happen," with varying degrees of likelihood.

Using the die example, you know that a die has 6 sides, and that the likelihood of any of the 6 sides landing upright are equal (not accounting for the thrower's skills, the game surface, the quality and weight of the die, mysterious gravitational disturbances in the room, etc.). Each side has a 1 in 6 chance of being rolled.

Using the second example of 100 tennis balls: if half the balls are red, the other half are yellow, they're mixed together, and you're blindfolded, what are the chances you'll pick up a red ball? Excluding all other factors, the answer is 1 in 2, and therefore you have an equal chance of pulling out a yellow ball or a red ball.

This illustrates a basic principle of **Probability**. Note that the combined sum of all possible outcomes in each case equals 1:

Die Example: 1/6 (chance of "1") + 1/6 (chance of "2") + 1/6 (chance of "3") + 1/6 (chance of "4") + 1/6 (chance of "5") + 1/6 (chance of "6") = 6/6 = 1.

Balls Example: + 1/2 (chance of "yellow") + 1/2 (chance of "red") = 2/2 = 1.

Fine, but What is Meant by "Practical Probability?"

Before we continue, though, let's be clear about something. We aren't going to learn how to compute the probability of an occurrence within the context of refined scientific experiments in this lesson. We're talking about practical situations, and therefore, you need to become comfortable with a certain level of anecdotal flexibility. Let's illustrate this with some examples.

What are the odds that your smartphone weather app's forecast for tomorrow is accurate, given that it was accurate in predicting the next day's weather on 15 of the last 20 days? The answer is 15/20, or .75.

What are the odds that your favorite author's next work will be over 200 pages long if all 5 of her previous books are far longer than this? The answer is 5/5, or 1.

After reading these examples, you may think to yourself, "Wait, this isn't right." You'd be correct!

In the weather example, there could now be a large storm system approaching the area that meteorologists are all but certain will result in precipitation, thus drastically increasing their chances of being correct. If this were the case, the odds that their forecasts are correct would likely be higher than 75%. The numbers don't necessarily model reality.

In the book example, you of course can't be certain that the author's next release will be over two hundred pages; she may switch gears entirely and put out an experimental 48-page work. Again, the numbers don't necessarily model reality. The *past* doesn't necessarily predict the future.

However, what if you didn't know about the impending storm? What if you didn't see the interview with the author where she hinted at the stylistic change? When we discussed **Guestimation**, you learned that you want to collect whatever data you can before making a decision. In the same way, *it is safer to make an assumption based on* <u>some</u> *information than it is to guess wildly.*

This is the essence of **Practical Probability**; use any available information to create the best possible answer. The answers you come up with aren't always going to be incredibly accurate, but you should strive to make them the best guess you can given the historical or available data. The numbers won't be exact, but you're better off than you would be had you done no math.

With that said, let's continue.

89

Basic Multi-Faceted Probability

Let's look at a scenario that involves multiple pieces of data.

Imagine you get a letter in the mail from your car's manufacturer alerting you that your car (a 2008 model, built between January 1, 2008 and December 31, 2008) could be recalled due to a mechanical issue. According to the company's records, the problem began occurring at all production facilities at the end of July. So, if your car was built after July 31, it will definitely be recalled (100% chance, or a **Probability** of 1).

This would mean that the chances of your car being recalled are 5/12 (42%), because there are 12 months in the year, and 5 of them occurred after July 31.

This is somewhat simple, so let's make it more interesting by assuming you know your car was built after April 30, but nothing more specific than that. This means it could have been built at any time between May 1 and December 31.

You know the likelihood of a recall if the car was built after July 31 (100%, or 1), and that it was definitely built after April 30. Now you need to compute the likelihood that the car was built *after July 31, given the fact that it was definitely built after April 30.*

As there are 8 months remaining in the year after April 30, and you're only concerned about those cars built after July 31 (when there are 5 months left in the year), there is a 5/8 (62.5%) chance that your car would therefore be recalled.

Work the following problem out on your own. If Peter is exactly 15 years old and you only know that Janice is less than 20 years old, what is the probability that Janice is older than Peter?

The answer is 5/20 or 1/4 (or 25%). Out of Janice's 20 potential ages (1 through 20), only 5 of them (16, 17, 18, 19, and 20) would make her older than Peter. 5/20 equals 1/4, which is 25%.

Don't overthink; calmly devise ways to translate word problems into math problems you can solve.

Simple "Or" Events

The type of simple **Probability** you just learned is fairly intuitive to most. Beyond this, many people tend to hold some strange and incorrect ideas.

In the die example, you know the chances of rolling a "2" (1/6) and you know the chances of rolling a "5" (1/6). What are the chances that you'll roll *either* a "2" *or* a "5?" Think about it: does it make sense for it to be more or less likely than 1/6? There are more outcomes you would consider a success now, aren't there? If chances of a successful outcome have increased, the probability must be higher. Don't overthink it or get scared off by the math—think about the problem logically.

Simply add the respective odds together. Here, the chance of rolling *either* a "2" *or* a "5" is 1/6 + 1/6, or 2/6. In other words, being happy with 2 of the 6 possible outcomes means that 1/3 of the possible outcomes will make you happy, as 2/6 = 1/3.

Apply this concept to the tennis ball example. If 50 of the 100 tennis balls are red (50 is 1/2 of the total ball count) and 50 of them are yellow (the other 1/2), what are the odds that *either* a red *or* a yellow ball will come out if you choose one without looking? When you add together the odds of yellow (1/2) + the odds of red (1/2), you find that the odds are 1. As you learned, a **Probability** of 1 means that the event is definitely going to occur. Since all the balls are *either* red *or* yellow, and either would make you happy, you know this to be true.

Practical Examples: When betting on a horse race involving eight horses with identical records and statistics, what are the odds that either Midnight Rider or Mississippi Queen will take first place?

Two horses out of 8 would be 1/8 + 1/8, which equals 2/8. Reduce 2/8 to 1/4, and the answer is 1/4 or 25%. There is a 25% chance that either Midnight Rider or Mississippi Queen will take first place.

Successive or Simultaneous Events

Sometimes events occur one after another, and sometimes they occur at the same time. Understanding **Probability** in either situation isn't too complicated.

Successive Events

To determine the likelihood of **Successive Events** occurring, multiply the likelihood of each event together. The product is a smaller number, which makes sense if you think about it; the odds of two **Successive Events** occurring would be slimmer than either one of the two events occurring independently. For instance, the odds of being struck by lightning *and then* becoming the victim of a shark attack are considerably less than *either* being struck by lightning *or* attacked by a shark.

Apply this to the die example: when rolling a single die twice in a row, what are the chances of rolling a 3 both times?

You know the likelihood of a 3 coming up *once* is 1/6, so to figure out the chances of it happening *twice* successively, multiply 1/6 × 1/6. The answer is 1/36.

> *Tip: In case you forgot (it's probably been a while), to multiply fractions, simply multiply the top numbers (numerators) and then the bottom numbers (denominators). If possible, simplify the resulting answer.*

What about the likelihood of *different* sides of the die coming up successively? What would be the likelihood of rolling a 3 on the first throw, followed by a 5 on the second? Don't let the change confuse you; keep your emotions and instincts out of the picture and simply do the math. Regardless of which sides of the die are involved, the likelihood of each isolated occurrence (rolling a 3 or rolling a 5, respectively) is 1/6, so the math is exactly the same: 1/6 × 1/6 = 1/36.

Simultaneous Events

The method for calculating the likelihood of **Successive Events** also works for **Simultaneous Events** (such as rolling two dice at the same time). Like **Successive Events**, the likelihood of two **Simultaneous Events** occurring is slimmer than either one of the two events occurring independently. To revisit the example used earlier, the odds of being struck by lightning *during* a shark attack are considerably less than *either* being struck by lightning *or* attacked by a shark.

By the same token, if you rolled two dice at the same time, the likelihood of a 3 and a 5 coming up at the same time are the equal to the odds of rolling a 3 *and then* a 5 on two successive rolls of the same die: $1/6 \times 1/6$, or $1/36$.

Let's look at a simple coin example. When flipped, a coin has a $1/2$ chance of landing heads-up and a $1/2$ chance of landing tails-up. Together, these add up to 1, because *something* will be facing up no matter what. When flipping three coins simultaneously, what are the odds that all three will land heads-up?

$1/2 \times 1/2 \times 1/2 = 1/8$, so the answer is $1/8$.

Think about your 100 tennis balls. There are 50 red and 50 yellow as before, but this time, 25 of the red and 25 of the yellow are marked with the letter "Q." Without looking, what are the odds of pulling out a red ball with a "Q" on it?

You know that there are 100 total balls, and only 25 of them are *both* red and marked with the "Q," making the odds $25/100$ or $1/4$. While this is simple enough to do without breaking it down, let's examine each possibility to give you practice; you'll be employing the same principles while solving much more complex problems. In this example, you would multiply the likelihood of pulling a red ball $(1/2)$ by the odds of pulling a "Q" ball $(1/2)$. Since $1/2 \times 1/2 = 1/4$, the answer is $1/4$.

Here's a more complex example. What would happen if you rolled a magic die six times, and after every roll, the side that landed face-up disappeared? So after the first roll, it becomes a 5-sided die, then a 4-sided die after the second roll, etc. Let's compute the likelihood of rolling a 1, then a 2, then a 3, then a 4, then a 5, and finally, a 6.

The likelihood of hitting a 1 on the first roll is 1/6 (the 1 then disappears). The likelihood of rolling a 2 on the next roll would be 1/5 (as there are now only 5 sides, 2 through 6). The likelihood of rolling a 3 on the third roll would be 1/4, the likelihood of rolling a 4 on the fourth roll would be 1/3, the likelihood of rolling a 5 on the fifth roll would be 1/2, and the likelihood of rolling a 6 on the sixth roll would be 1/1, because it would be the only side left (just ignore the fact that it's hard to imagine a one-sided die). So the likelihood of a 1-2-3-4-5-6 sequence occurring is $1/6 \times 1/5 \times 1/4 \times 1/3 \times 1/2 \times 1/1$, which equals 1/720, or 0.14%. In short, it's probably not going to happen.

Practical Examples: When betting on a horse race involving eight horses with identical records and statistics, what is the likelihood that Midnight Rider will take first place and Mississippi Queen will take second (an "exacta" for you gamblers out there)?

This is an example of **Simultaneous Events**. There is a 1/8 chance that Midnight Rider will take first place and a 1/8 chance that Mississippi Queen will take second. When multiplied, 1/8 and 1/8 equals 1/16.

When flipping three coins into the air simultaneously, what is the likelihood they will come up heads, tails, and tails, respectively?

Since there are three events with two equally possible outcomes each, the likelihood is 1/2 × 1/2 × 1/2, or 1/8. Again, don't overthink it.; use logic—not instinct—and stay calm.

In each of these examples, please note that the answer is computed as "the likelihood of each event occurring independently, multiplied together."

"Or" Events with Duplicates

Let's go back to the 100 tennis balls example (50 red, 50 yellow, and 25 of each color marked with a "Q"), but this time, you want to know the likelihood of pulling a ball that is *either* red *or* marked with a "Q."

First, take the likelihood of pulling a red ball (50/100 or 1/2) and add that to the likelihood of pulling a marked ball (50/100 or 1/2). You *add* here because this is an "or" problem ("or" problems require addition; "and" problems require multiplication). When you add 1/2 and 1/2 together, you get 1, which means in **Probability** terms that your arbitrarily selected ball will definitely be either red or marked with a "Q."

Wait a minute, that's not right! Isn't there a glaring problem with this answer? You want to choose balls that are *either* red *or* marked. Sure, yellow marked balls fit the criteria, and so do red unmarked balls; however, yellow unmarked balls won't work. How can the answer be 1—meaning there is a 100% chance that an arbitrary selection will produce the desired result—if there exists a subset of the balls that won't work?

For now, take your answer so far, 1, and call it the **"Or" Subtotal**.

The problem lies in the group of 25 balls that are *both* red *and* marked. You counted them twice: once in the group of 50 red balls, and again in the group of 50 marked balls. To determine the correct probability, these duplicate occurrences need to be eliminated, so the final step is to account for the likelihood of coming upon each attribute independently. To do so, multiply the likelihood of pulling a red ball (1/2) by the likelihood of pulling a marked ball (1/2). $1/2 \times 1/2 = 1/4$. This is called the **Duplicate Set**.

Subtract this **Duplicate Set** from your **"Or" Subtotal**. 1 (the **"Or" Subtotal**) – 1/4 (the **Duplicate Set**) = 3/4 (final answer).

A Few Problems to Get Started

Let's look at a few **Probability** problems and determine which of the methods you've learned so far would be best for each. Remember: don't overthink it, but remain calm and confident, and let logic prevail. Once you've solved them all, check your solutions against the answers provided.

1. Desmond is a deranged, self-destructive lunatic and patron of an illegal underground gambling compound. Surrounded by the distinct sounds of cockfights and brawls, he gets involved in a high-stakes game of Russian roulette. He is given a revolver (a handgun that holds up to six bullets), and as per the usual rules of the game, this revolver is loaded with a single bullet. He has to spin the cylinder (to somewhat randomize the outcome) and pull the trigger with the gun pointed at his temple. If Desmond

survives the first round, he spins the cylinder again (again randomizing the outcome) and tries a second time. If he survives both rounds, he stands to win 376,116 Thai baht (roughly $12,000 USD). **What type of problem is this? What is the likelihood that he will survive?**

2. A new ice cream shop called The Cream-atorium opens in a nearby town. Unlike most ice cream joints, this one has committed to a horror theme, and features black walls, fake blood, gory zombie mannequins, crypts, and realistic-looking spider webs. At a subsequent town hall meeting, locals complain, saying the shop is "inappropriate," "godless," and "in poor taste." After calming everyone down, the meeting organizer promises to take a poll of arbitrarily-chosen residents. The next week, he gathers a group of 200 strangers in the town square. He asks, "Who likes ice cream?" and 120 people raise their hands. He then asks, "Who likes horror movies?" 60 people raise their hands. He addresses the angry crowd at the next town hall meeting and says, "My poll shows that 180 out of 200 people are into either ice cream or horror." **There's an obvious flaw in his logic; what is it? What type of problem is this?**

3. Sean is into Rock Paper Scissors, or "RPS," as the pros say. He's practicing for a big tournament overseas, and as all pros know, being able to read an opponent's physical cues is key to success. When his practice partner falls ill and fails to show for several weeks, Sean is forced to train with his young sister. This poses a problem, as amateurs are harder to read than pros; unencumbered by the "mind game," they are unpredictable and wild, choosing weapons arbitrarily and with little regard for strategy. Sean decides that the best approach against his sister is to make wild, arbitrary choices as well. For their first match,

Sean and his sister both shoot "rocks." **Imagining that both Sean and his sister's selections are truly random, what are the chances of this "rocks-rocks" outcome occurring? What type of problem is this?**

4. Tanya, an accountant, is walking through Central Park when a street performer lures her over to a table. He places a small ball under one of three identical cups, challenging her to follow the ball through a series of fast transfers. Being a sharp, detail-oriented person, she accepts his challenge and bets a dollar on it. Halfway through the game, a dog barks in the distance, followed by the sound of breaking glass. Tanya's eyes momentarily dart away from the play area. When she looks back, she's annoyed to find she's lost track of the ball. At the end of the game, Tanya has no idea which cup the ball is under. She arbitrarily chooses the middle cup. **What type of problem is this? What is the likelihood that she's guessed correctly?**

5. Janet is a superficial music fan who only really cares about hanging out with rock stars. At a concert, she recognizes a musician (Jim "Rattlesnake" Bonnar, from the band *Eroding Codpiece*) and walks up to him. "Hey there!" she exclaims, "Can I *please* get a backstage pass? I'm your biggest fan!" Familiar with her type, the musician points to a band mate and says, "Sure, I'll give you a pass. But first, you have to tell me which instrument either he or I play in the band." Janet thinks for a moment, twirling a lock of her hair with a blank expression, and realizes that her bluff has failed; she truly has no idea. But she does know that the band has four members: a guitarist, a singer, a bassist, and a drummer. She'll just have to guess. **What type of problem is this? What are the odds that she'll guess either of the two correctly?**

Answers:

1. This is a **Successive Events** problem. To solve the problem, you must multiply the likelihood that Desmond will survive each individual round of the game. Note carefully the wording of the problem: you aren't looking for the likelihood that he will catch a bullet ($1/6 \times 1/6$), but rather the likelihood that he *won't* ($5/6 \times 5/6$). The answer is therefore 25/36; he has about a 70% chance of survival (or a **Probability** of 0.7). Not bad for $12,000. Good luck, man!

2. This is a sneaky problem—despite the organizer's use of the word "either," this is an "or" with duplicates problem. In our vocabulary of **Probability**, "either" denotes exclusivity, whereas "or" denotes multiple possibilities. The pollster's mistake was that he didn't specify that each person can only cast one vote, resulting in substantial overlap. As a result, while it appeared that 180 (120 + 60) out of the 200 people would be into the concept of the ice cream shop, in reality, some of those people most likely enjoy *both* ice cream and horror (let's say 40 for the sake of the example), and you need to account for these people. Add the outcomes of the two polls (120 + 60 = 180, your **"Or" Subtotal**) and subtract 40, the **Duplicate Set**. 180 − 40 = 140, This means that 140/200 or 7/10 of the people polled enjoy ice cream *or* horror (though not necessarily exclusively). So, there is a 70% chance—a probability of 0.7—that someone from the polled group would be interested in *at least one* aspect of the Cream-atorium. The members of the **Duplicate Set** are surely the shop's ideal target demographic.

3. This is a **Simultaneous Events** problem. To solve it, you must multiply the odds of the two contestants shooting "rocks"

independently, which is 1/3 × 1/3; thus, the odds of both Sean and his sister shooting "rocks" simultaneously is 1/9.

4.   This is a basic **Probability** problem. Don't overthink it just because it's positioned after three more complex problems. You simply need to divide the number of guesses Tanya can make (1) by the number of options she has (3 cups) to arrive at the answer, 1/3. She has a 33.3% chance of guessing correctly.

5.   On the surface, this is a simple "or" problem, since Janet doesn't need to be right about *both* band members' instruments, just one of them. You need to add the likelihood of each of her individual guesses being correct. Janet has a 1/4 chance of guessing correctly about Musician A's role, and a 1/4 chance of being right about Musician B's. Hence, she has a 2/4 (or 50%) chance of getting her backstage pass. However, before you finish, let's play with this example a bit more. This problem doesn't exist in a vacuum. Janet most likely would not guess that Musicians A and B are *both* drummers, since there is only one drummer in the band. So, she'd have a 1/4 chance of guessing Musician A's role correctly, but for Musician B, she'd most likely have eliminated the instrument she chose for Musician A. Does this affect the **Probability** of her being correct about Musician B? Think about it. Assuming her guess for Musician A doesn't result in any feedback about whether or not she was correct, the selective elimination of an option doesn't actually shrink her set of possible answers. She still has a 1/4 chance of guessing Musician B's instrument correctly, as long as no additional information becomes available.

Hopefully, you learned a thing or two about **Probability** from this lesson. If you can commit these principles to memory, you will be able to decode most of the practical probability problems you should encounter in your personal or professional life.

These concepts aside, the most important practical thing to know about **Probability** is that your instincts usually fail to recognize the actual likelihood of events occurring (especially rare or sensational events). It's human nature. For instance, the odds of getting a royal flush in a hand of poker are 649,740 to 1. Did you think it was that rare? Your odds of becoming a victim of violent crime are slimmer than you might think. The same can certainly be said of your chances of being attacked by a shark or struck by lightning. Simply remembering and acknowledging this will serve you well.

Quick Review

1. **Probability**: **Probability** is the likelihood that an event will occur, assuming that you don't know for certain whether or not the event will *definitely* occur. **Probability** is measured on a scale of 0 ("not happening") to 1 ("definitely happening").

2. **Practical Probability**: **Practical Probability** is the art of applying **Probability** principles—often somewhat loosely—to everyday situations.

3. **Basic Probability**: When computing basic **Probability**, you are simply comparing the likelihood of an event occurring with the likelihood of other events occurring (given that one such event must occur).

4. **Simple "Or" Events**: When computing a simple "or" event, you must consider the likelihood of *any* (but not necessarily *all*) of two or more events occurring, so you add the occurrences' individual likelihoods together.

5. **"Or" Events with Duplicates**: Some "or" events contain duplicates. In other words, calculating **Probability** as if it were a pure "or" problem would double-count the likelihood of certain outcomes. After adding the individual events' **Probabilities** together (an **"Or" Subtotal**), you must then subtract the duplicate events (called the **Duplicate Set**).

6. **Successive or Simultaneous Events**: **Successive Events** and **Simultaneous Events** are events that are related in that they must all occur to meet your criteria of success. To compute the likelihood of two (or more) **Successive** or **Simultaneous** events occurring, multiply the likelihood of each individual event occurring independently.

Lesson 3: The Math Behind Options

Every new moment presents you with a new suite of options. You have immense control over the order in which you perform tasks and otherwise organize your life. You accept this in a large-scale, abstract way, but do you understand the math that informs these options?

If you have two objects or events—A and B—you know that there are two ways they can be ordered: A followed by B, or B followed by A. And you know that if you can only choose one, there are two options: A or B. But how many ways are there to order a list of sixteen options? How many combinations could you make from these sixteen options if you could only choose five at a time?

To understand the math behind an option is to better understand its scale, context, and potential impact. It informs your decision-making freedom. It gives you more control, which may give you a greater degree of confidence in your decisions.

Let's get right into it and address two common and useful applications of this concept.

Determining the Number of Possible Ordering Options

You wake up on a Sunday morning and mentally review your agenda for the day. You need to take a shower, walk the dog, get gas for the car, go for a jog, pick up groceries, and change the batteries in your smoke detector. You may perform these chores in any order you wish, but you must do them all before nightfall. Naturally, you must consider the order in which you'd prefer to perform these tasks. After a moment of pondering, you realize that there are quite a few options.

"I could walk the dog, then take a shower, fuel up the car, go for a jog, pick up the groceries, and change the batteries; wait, I should jog *before* I shower. Right. So I could go for a jog, take a shower, walk the dog, change the batteries, pick up groceries, and then get fuel on the way home. But I'm pretty low on gas..."

There are six tasks to be performed. All in all, how many choices do you have? Without getting into any calculations, take a guess.

The method by which such problems are computed is actually somewhat simple, but most people are unaware of how to even begin. Whenever you have an array of options and you'd like to know how many possible arrangements there are for them, take the number of options (here, it's 6) and multiply this number by *one number less* (5), then by *one number less* (4), and so on, until you reach 1. In this example, $6 \times 5 \times 4 \times 3 \times 2 \times 1 = 720$. The formal mathematical term for this is "factorial" (one would say, "6 factorial is 720").

Of course, you can ignore the part where you multiply by 1, as multiplying any number by 1 results in the original number.

Most people would guess that there were significantly fewer than 720 ways to order 6 tasks. How close was your guess? If you take only one concept away from this lesson, let it be this: *there are usually far more options than you would guess from instinct alone.*

Determining the Number of Options Given Limitation

You are moving from a house with a small cupboard to one with an even smaller cupboard. You currently own three coffee mugs (let's call them Mug A, Mug B, and Mug C). After moving, you will only have space to store two. You must dispose of one mug and take two with you. How many different choices do you have?

Math is hardly needed; you can take A and B, B and C, or A and C, giving you three choices. The order of each choice doesn't matter, as "A and C" is the same as "C and A." This is easy to figure out with a small set of easily-imagined options, but how do you handle large, complex problems of a similar type?

You must first determine two numbers (we'll call them the "bottom" and "top" numbers), then divide them. Let's apply this to the simple mug example (since you already know the answer is 3). Don't worry if you don't quite follow along at first, as we'll go over it in more detail shortly.

Determining the "bottom" number is the most difficult part, so get it out of the way:

1. Take the total number of mug options (3) and subtract from it the number you're allowed to keep (we'll call this the **Limitation Set**). The **Limitation Set** is 2, as the new cupboard will only hold 2 mugs. 3 (total options) − 2 (**Limitation Set**) = 1.
2. Next, compute the factorial of this number; in this case, the factorial of 1 is 1.
3. Lastly, multiply step 2's answer by the factorial of the **Limitation Set**. 1 (from step 2) × 2 (2 factorial) = 2.

The "bottom" value is 2. Hold onto that number.

The "top" number is the factorial of the total number of available options. You have three options (mug A, mug B, or mug C), so 3 × 2 × 1 = 6.

You then simply divide the top number by the bottom number. The top (6) divided by the bottom (2) is 3; this is the number of choices. You know this is true, because you first solved the problem on your own using simple logic.

You'll soon try this method out on a problem that you *wouldn't* be able to solve using your intuition. But before doing that, let's review the formula and look at this method in more detail.

Factorial is represented by a *!*. If *T* is the total number of options (3 mugs) and *L* is the **Limitation Set** (the 2 mugs you can keep), then your number of choices (*C*) is:

$$C = \frac{T!}{(T - L)! \times L!}$$

Memorize this. It looks complicated at first, but it becomes simple with a little practice.

Here's a more difficult example. Let's say you're going to visit a friend overseas, and you have twelve 10-lb items you want to bring: a folding tent, a bag of toiletries, a bunch of clothes, a box of American candy (a gift for your friend's mother), a few English-language books, a large gift for your friend, a metal detector (your latest hobby), a spool of wire (you never know when you'll need wire), a geology toolkit (your other latest hobby), a well-padded camera and several lenses (you have too many hobbies), and…well, two other items. You are all packed, but in reviewing your flight itinerary, you notice that there's a 50 lb. limit on luggage. Out of the 12 items, you need to choose 5. How many possible combinations can you make?

First figure out the "bottom" number. Take the total number of options to choose from (12), subtract the **Limitation Set** (5), and compute the factorial of this number (7 factorial = $7 \times 6 \times 5 \times 4 \times 3 \times 2 \times 1$), which gives you 5,040. Then multiply this number by the factorial of the **Limitation Set** (5 factorial = $5 \times 4 \times 3 \times 2 \times 1 = 120$), to arrive at 604,800 ($5,040 \times 120 = 604,800$). Hold onto the "bottom" number.

$$C = \frac{T!}{(12 - 5)! \times 5!}$$

Next, determine the "top" number. Take the total number of options to choose from (in this case, 12), and compute its factorial ($12 \times 11 \times 10 \times 9 \times 8 \times 7 \times 6 \times 5 \times 4 \times 3 \times 2 \times 1$) to arrive at 479,001,600.

$$C = \frac{12!}{(12 - 5)! \times 5!}$$

Divide the "top" number by the "bottom" number.
479,001,600/604,800 = 792. There are 792 possible five-item
combinations you could build from the original twelve items. That's
quite a few. You may want to leave the spool of wire home, after all.

$$792 = \frac{12!}{(12-5)! \times 5!}$$

A Few Problems to Get You Started

Let's look at two problems. The answers are listed below.

1. You're playing a riveting game of Scrabble with your great
   aunt, and while she's taking ages to decide on her next word,
   you examine your letters. You have P, N, I, N, G, C, and A.
   Trying to come up with a good word for your next turn, you
   begin arbitrarily reordering them and realize that there are
   more possible arrangements than you thought there would be.
   **What type of problem is this, and how many different
   possible arrangements are there?**
2. After guessing the correct number of gumballs in a jar at the
   state fair (see **Guestimation**), you win a $10 gift card to *Frank's
   Bottom Dollar Discount Emporium*, where everything is $5. You
   have your eye on six back-to-school items for your oldest child.
   Obviously, you can only choose two (2 × $5 = $10) without
   exceeding the value of the gift card. **What type of problem is
   this, and how many different possible combinations are
   there?**

Answers:

1. This is a "Determining the Number of Possible Ordering Options" problem. Since there are 7 letters, the total number of possible arrangements is 7 factorial (7 × 6 × 5 × 4 × 3 × 2 × 1), or 5,040. Deceivingly high, isn't it?

2. This is a "Determining the Number of Options Given Limitation" problem. The formula is $T!/[(T - L)! \times L!]$, so once you plug in the values, it's $6!/[(6 - 2)! \times 2!]$. This is further broken down to $720/(24 \times 2)$, or $720/48$. Since $720/48$ equals 15, there are 15 different two-item combinations that you could make out of the 6 options you're considering.

Realistically, you could spend a lifetime studying the mathematics behind options, but the applications you just learned are those that tend to come up most often. You can of course research the subject further if you wish.

Quick Review

1. **The Math Behind Options**: To understand the math behind an option is to better understand its scale, context, and potential impact; it informs your decision-making freedom.

2. **"Determining the Number of Possible Ordering Options" Problems**: To determine the number of possible ways to order a set of objects, tasks, or events, take the total number of items and calculate its "factorial:" multiply the number by "itself minus 1," then by "itself minus 2," and so on, until you reach the number 1. For example, "4 factorial" is equal to 4 × 3 × 2 × 1, which is 24.

3. **"Determining the Number of Options Given Limitation" Problems**: To determine your possible choices when

considering which items (a **Limitation Set**) you'd like to select from a larger set of items, use the formula $C = T!/[(T − L)! \times L!]$. In this formula, C represents your choices, T represents the total number of options, L represents the **Limitation Set**, and $!$ means "factorial."

Lesson 4: Quick Percentages

"GO DOWN DEEP ENOUGH INTO ANYTHING AND YOU WILL FIND MATHEMATICS."

—CHARLES SCHLICTER, EDUCATOR

Schlicter is surely correct; mathematics is the foundation for both the concrete and abstract sciences that attempt to explain the nature of the universe. But we're not here to wonder at the beauty of mathematics; in this book, the focus is on practical applications, and becoming comfortable with percentages is about as practical as it gets.

People tend to think of things in terms of portions of 100 (even those who live in the backward, non-metric confines of the United States[1]). Most of us can understand that "100% of something" is representative of "the whole" of that thing and have learned to envision parts of a whole based on a percentage.

It's easy to imagine 50% of a pie, a glass 25% full, a water cooler that's 80% empty, and a novel that's 50% read; however, can you so

----

[1] Although this issue had technically been addressed by the Metric Conversion Act of 1975, the forgiving nature of the text allowed the existing system to prevail indefinitely.

readily imagine a novel in which 378 out of 756 pages have been read? Life doesn't always hand you information as neatly packaged percentages; 378/756 is the same as 50%, but the way it's conveyed makes it more difficult to quickly understand.

Let's talk about how to translate somewhat complex information into percentages.

24% of People (1 in 6) are Bad with Percentages

That was a joke. Get it? 1/6 is actually 16.667 percent. See...eh...never mind, you'll get it later.

I'll never forget the day. It was summer in the early 2000's, and I was working at a company that produced FDA submission software for pharmaceutical companies (exactly as exciting as it sounds). I was in a meeting in a large boardroom. As usual, I was staring out the window, battling boredom with vivid jetpack fantasies. The following exchange unfolded before me:

Heavyset Guy: "Well, the [something something] went down from 520 to about 330 this month alone."

Hairy-Armed Guy: "Hmm. Almost a 40% loss. What's causing it?"

I snapped out of it. How had Hairy-Armed Guy performed that calculation so quickly? I worked out the arithmetic on a piece of paper, and sure enough, he was correct. I'd seen this before, when people seemingly pull these figures out of thin air, as though it was as simple as adding two and four.

By this point in the book, you shouldn't be surprised to learn that it's neither magic nor evidence of some innate savantism. The trick is to understand *how* to solve this type of problem and then address each problem calmly and rationally. Let's figure out what these rain men are doing that you aren't (spoiler: they're using shortcuts).

The Race to 100

Earlier, we highlighted the difficulty inherent in imagining a book that's 378/756 read versus one that's 50% read. The latter exists within a familiar framework; it explains the problem as a part of 100 percent. The former does not. One way to understand the former is to convert the denominator (756) to 100 and alter the numerator (378) proportionally, to retain the relationship between the two numbers.

Let's take a deeper look at the exchange between Heavyset Guy and Hairy-Armed Guy. In this problem, we're calculating the percentage lost when 520 dropped to 330. First, understand that 520 is the denominator and is therefore the value that needs to correspond to 100%.

The simplest method is to round both numbers and multiply the numerator (330) by the denominator's relationship to 100. This second step can require some creativity and comfort with numbers. Here, 330/520 would round to 3/5, and because 5 goes into 100 exactly 20 times, you would multiply the numerator (3) by 20 to arrive at 60. Therefore, 330 is roughly 60% of 520. The real answer is 63.5%, so this method is not particularly precise, but it produces answers that suffice in most conversational or practical contexts. Since $100 - 60 = 40$, Hairy-Armed Guy was right about the loss.

When using this simple, rough method, you want to do a few things before considering the denominator's relationship to 100:

- Round at a scale that makes sense. With the problem 711/1,294,244, you wouldn't want to round both numbers to the nearest 10 (710/1,294,240) or 100 (700/1,294,200). With a denominator as large as 1,294,244, you can round to the nearest 1,000; 10,000; or even 100,000; which still usually gets you close enough to the answer. Rounding this example to the nearest 100,000 gives you 0/1,300,000, which equals zero; the actual answer is 0.0005.

- Round in the direction of the closest round number given said scale. This may seem obvious, but care here will improve the precision of the answer. With three-digit numbers, if a number is in the "top" of a hundred-number span (X50 – X99), you'd round up. If it resides in the bottom (X01 – X50), you'd round down. For six-digit numbers, X00,000 through X50,000 would round down and X50,000 through X99,999 would round up. 44/81 would become 40/80, 39/109 would become 40/110, and 23,689/73,938 would become 24,000/74,000.

- Simplify as much as you can. 40/110 and 4/11 both equal 36.4%. 24,000/74,000 and 24/74 both equal 32.4%.

Let's look at some examples.

Here's an easy one. 54/83 would round to 50/80 and simplify to 5/8. 8 goes into 100 a little more than 12 times, and 12 × 5 = 60. The answer should therefore be a little more than 60, maybe around 63%. 54/83 is actually 65%, so this is pretty close for having done very little math. The more comfortable you become with math as a language, the easier these types of conversions will become.

Let's end with the novel example to illustrate the amount of precision that's achieved at different levels of detail. 378/756 translates to a clean 50%, but it's deceiving at first glance.

First turn the three-digit numbers into single-digit numbers. Convert 378 to 4 and 756 to 7. You know that 7 goes into 10 a bit less than 1.5 times, and 1.5 times 3 equals 4.5, so the answer produced by the simplest method is a bit under 45%. Let's say 42%.

Now, let's solve the same problem with each number converted to two digits. 378 becomes 38 and 756 becomes 76. You know that 76 is close to 75 and that 75 is ¾ of 100, so it's safe to say that 76 is about ¾ of 100. So, ¾ is the key. You then ask yourself what number 38 roughly 3/4 of? A third of 38 is about 12 (12 × 3 is 36), so 38 is roughly ¾ of 38 + 12, or 50. Your answer would therefore be about 50%, which is spot-on.

As you can see, this is much more of an art than a science and simply requires you to be comfortable enough with math to make these associations.

When Exact Percentages are Required

These methods are useful when extreme precision is not required. Imagine that you write a children's book and then poll strangers in a local park on whether or not they like the book cover. In this instance, you probably don't need to know that 81.8% of those polled liked it—only that around 80% did.

However, if the situation calls for exact answers with decimal places (such as scientific applications), these methods will not suffice. Most of the time, life doesn't require scientific precision, and when it does, it's often best left to tools such as calculators. However, if you're interested, you'll soon learn strategies that allow you to perform arithmetic with surgical precision. This of course requires a bit more time and brainpower, so use the rough methods whenever it's reasonable to do so.

A Few Problems to Get You Started

Convert the following fractions into percentages in your head. Work quickly and remember that your answers should fall within a reasonable margin of error. The actual percentages are below.

1. 48/360
2. 27/120
3. 910/1033
4. 7/62
5. 1,280/4,999
6. 88,519/102,444

Answers:

1. 13.3%
2. 22.5%
3. 88.1%
4. 11.3%
5. 25.6%
6. 86.4%

Quick Review

1. **Quick Percentages**: To translate a fraction into a percentage, first round the fraction, taking into account the level of precision required by the situation. Simplify the fraction, ideally ending up with rounded one- or two-digit numbers. Identify the relationship between the denominator and 100, and apply that relationship to the numerator. This method is very fast but has the potential to produce somewhat imprecise answers.

Lesson 5: Value Comparison

Though value comparison is one of math's most practical life applications, it remains a mystery to most of us.

The need to compare two or more values comes up on a daily basis. Like any mathematical challenge, if you don't have a method, you have to "wing it," which can produce grossly inaccurate results (especially when conversion is required).

There are several types of **Value Comparison**.

Simple Value Comparison

These two examples will illustrate how to perform **Simple Value Comparison**.

Example 1: Which is the better deal: $8.49 for four Pee Wee Herman Christmas tree ornaments or $9.99 for six? Although you want as many as possible (and who wouldn't?), imagine you're more concerned with value than sheer quantity. This becomes a simple question of the cost per unit (the individual ornaments) in each of the two scenarios.

Example 2: You work for a movie theater, and can either purchase 934 pairs of disposable 3D glasses for $418, or 1,020 pairs for $490. Which is the better deal?

In most practical cases, the larger the problem's scale, the less likely it is that a comparative difference will come down to a tiny margin. Much like fraction-to-percentage translation, you can often use rough approximations and forgo precise calculations. This is even easier when it comes to money, as retailers tend to use prices ending in .99 or .49, which are easily rounded to wholes and halves.

The first problem (Pee Wee ornaments) is easy. First, translate the word problem into a math problem by setting the price as the numerator and the number of units as the denominator.

- $8.49/4
- $9.99/6

Then, do some basic rounding; if the problem permits, reduce each number to two or fewer digits.

- $8.49/4 = $8.5/4
- $9.99/6 = $10/6

Finally, play around with the numbers until you arrive at a value per unit.

- $8.5/4 = ?: 4 goes into 8 twice and 0.5 is 1/8 of 4. The answer is $2⅛ or $2.12.
- $10/6 = ?: 6 goes into 10 once, and the remaining 4/6 equals 2/3. The answer is $1⅔ or $1.67.

The significant difference between the two values shows you that precise calculations would not necessarily have been more beneficial. In fact, the values above are similar to the results you'd get from a precise calculation.

Note that in solving this problem, 2⅛ turned into 2.12 and 1⅔ turned into 1.66. If you don't know how to translate fractions into decimals, don't worry; it will be covered soon in a lesson entitled *The F2D Method*.

To solve the second problem (3D glasses), first simplify the fractions:

- $418/934 = $420/930 = $42/93
- $490/1,020 = $490/1,000 = $49/100

Next, manipulate the figures to extract decimal numbers. Be creative!

- $42/93 = ?: Since 42 is roughly half of 93, you can begin with $0.5. Since 42 × 2 actually equals 84, you need to account for the difference between 84 and 93, which is 9. So, you know the actual answer will be slightly less than $0.5; specifically, it will be 9/93—or roughly 10%—less. $0.5 − $0.05 (10% of $0.5) = $0.45.

- $49/100 = ?: This is easy; 49/100 is $0.49.

It turns out that $418/934 breaks down to about $0.45/each, and $490/1,020 breaks down to about $0.49/each. The actual answers are $0.45 and $0.48, respectively. Again, rough work resulted in answers close to the precise ones.

Circumstances permitting, you can perform even faster and less precise calculations. Let's drastically simplify these problems by reducing each number to ten or less and see what happens.

The first problem:

- $8.49/4 = 8/4 = $2/each; the real answer is a bit higher since you rounded $8.49 down to 8.
- $9.99/6 = 10/6 = $1⅔ or $1.33.

The differences between these answers and the actual answers are $0.12 and $0.33, respectively. Even the least precise version of this calculation—with a considerable margin of error—results in values that clearly illustrate the comparison and let you know which option is a better value. If you need to know which of two choices is a better value, and you don't need to quantify exactly how much of a difference there is between them, you can perform these types of quick, imprecise calculations.

The second problem can be solved like this:

- $418/934 = $4/9 = a bit less than $0.50/each

- $490/1,020 = $5/10 = $0.50/each

Even with very loose rounding, it's apparent that the first option is a slightly better value.

There are no steadfast rules here; use your head, assess the situation's need for precision, and work through the problem using creativity and control, allowing for the comparison of two values using a like unit.

Work through the following problem: Who is more efficient (a better rate of speed): a data entry worker who enters 12 medical records per minute, or one who enters 48 in five minutes?

Answer: The first individual is faster; the first enters 60 records per five minutes (12 records per minute), and the second enters 48 per five minutes (9.6 records per minute).

Multi-Faceted Value Comparison

When comparing two or more options, you may have to consider additional factors.

Value Comparison Involving an Additional Operation

Imagine you are considering the purchase of a life-sized Hulk Hogan cardboard cutout from one of two different retailers. The first offers the item for $79.99 + $5.80 shipping and handling (S&H). The second retailer sells the same item for just $54; but because they're overseas, the shipping comes to $26. Which is the better deal?

Simply add the two options' prices and shipping costs together...

- $79.99 + $5.80 = $85.79
- $54 + $26 = $80

...and compare them ($85.79 > $80). This is somewhat obvious, but it illustrates the importance of organizing a problem before performing any calculations.

Let's look at another example:

You can get 180 dry-erase markers for $24.99 + $4.54 S&H, or 211 of the same markers for $19.99 + $7.80 S&H. How do you handle this?

A third factor has been introduced; while the first example involved comparing the costs for a single item, this time there are different quantities to think about.

Don't guess or panic. Take the time to organize the problem before doing any math.

Solving this type of problem is a two-step process. First add the prices and shipping to arrive at two totals ($29.53 and $27.79 if precise, or $30 and $28 if rounded). Second, use the **Value Comparison** calculations you just learned: $29.53/180 vs. $27.79/211 (or round numbers as you see fit).

You arrive at around $0.16/marker vs. $0.13/marker for both the precise and rounded calculations.

Work through the following problem. You can buy three live wolves for $2,480, but it will cost $180 in fuel and tolls to go pick them up. Or, you can purchase two for $1,531, but they'll be shipped to you free of charge. Assuming the unit is price-per-wolf, all costs considered, which is the better deal?

Answer: It's the second option. With pick-up costs, the first option comes out to about $866.67 per wolf. The second option comes out to $765.50 each. Even using drastically rounded numbers, the difference in value should be clear after some simple calculations.

Value Comparison Involving Unit Conversion

The first rule of type/unit conversion (example: 6.2 quarts vs. 28 cups) is—you guessed it—to avoid "winging it" at all costs. Resist your instincts and the need to deliver an answer swiftly. Use your mind. Remember to organize the problem before doing any math. Type/unit conversion problems require the additional step of converting one unit to the other.

In a stable market, Beluga caviar (wild sturgeon eggs) is often the most expensive publicly available food product; it costs around $6,500 per kilogram. You have been tasked with the purchase of a kilo of this caviar. After speaking with some specialty importers, the options are:

1. **Riviera Purveyors**: £4,000 per kilogram, plus £122 for refrigeration-guaranteed shipping and handling.
2. **Rich Dude Distribution**: $6,000 per 24-ounce container, plus $88 for refrigeration-guaranteed shipping and handling.

These options can only be compared once you convert one of the two monetary units and one of the two measurement units to match those of the other distributor. Given the expensive nature of this endeavor, you are going to use exact calculations.

Performing some quick online conversions reveals that £4,000 currently equals $6,428 (though in real life, this changes minute-by-minute), and £122 equals $196. The total cost for Riviera Purveyor's product and shipping is $6,624. The total cost for Rich Dude Distribution's product and shipping is $6,088.

However, you don't know if the quantities are comparable. A quick Internet search reveals that one kilogram equals 35.27 ounces, so Riviera is offering 35.27 ounces for $6,624; Rich Dude is offering 24 ounces for $6,088. This comes to $187.81/oz. vs. $253.67/oz., respectively. Riviera has the better deal.

Work through the following problem. Doug is 5'8" tall and weighs 228 lbs. He has entered a judo tournament in the "open" division, meaning his opponent can be of any size. When the matches are posted, he discovers he's pitted against a European fellow named Magnus, who is 180 cm. tall and weighs 99.7 kg. Doug wants to know who has a lower weight-to-height ratio.

Answer: Magnus has a lower ratio. Doug is 5'8", which converts to 172.72 cm., and 228 lbs., which converts to 103.42 kg. Doug's 103/172 vs. Magnus's 100/180 reduces to 0.6 kilos per cm. vs. 0.55 kilos per cm.

The takeaway is simple and universal; organize problems before attempting to solve them. Convert units of measurement to be uniform across the problem and be sure to take all facets into consideration.

Quick Review

1. **Value Comparison**: **Value Comparison** is the oft-required skill of assessing and comparing the relative values of two or more like things. Varying levels of precision and speed can be used depending on the situation. Often, reducing the problem to simpler terms only marginally impacts the answer's precision.

2. **Simple Value Comparison**: **Simple Value Comparison** involves the comparison of two or more things with a shared unit of measurement.

3. **Multi-Faceted Value Comparison**: **Multi-Faceted Value Comparison** refers to problems requiring additional operations or unit conversion before two or more things can be compared properly.

Lesson 6: Sorting

Years ago, there were many times when my CD collection became totally disorganized. I had the regrettable habit of haphazardly stuffing a CD into the nearest empty case and then jamming the case into the most convenient gap in the stack. The eventual result was total bedlam, and I wouldn't be able to find anything in a reasonable amount of time.

Eventually, the painful time for reorganizing would come. With a sigh, I'd pull all the cases off the shelf, open a case, reunite the disk inside with its proper case, and repeat.

Regardless of my habit of quickly disrupting the order, my theoretical organizational system had two levels. The collection was primarily organized by band/artist name. First came artists whose names began with A through Z, and then came artists whose names began with numbers. At the end of the collection came compilation disks (again in A-Z order by title). Within each individual artist's collection, albums would be **Sorted** in chronological release order, from earliest to latest.

What was the best way to take piles upon piles of disorganized CDs and place them into the order described? This question highlights the fact that most people go through life having never learned to **Sort** properly.

The study of **Sorting** algorithms is an important aspect of computer science. Computers often need to **Sort** enormous datasets; when there are millions or billions of items involved, efficient methods can save time and money. Many computer science **Sorting** principles can be useful for **Manual Sorting** ("manual" meaning *without the use of third-party tools*), but some of the most efficient **Manual Sorting** systems combine computer science principles with an array of human-specific factors.

Solving Simple Manual Sorting Problems

Let's begin with a simple **Sorting** scenario. Imagine you have a set of index cards that are printed with the letters "A" through "Z" (one letter per card). They're tossed in the air and you must put the resulting pile in order. This is a one-dimensional problem because the **Sorting** is based upon a single criterion: the letter. Assuming you'd like to **Sort** the cards in order from "A" to "Z," how would you begin?

The Insertion Sort

Though you may not be familiar with the name, this is usually the first **Sorting** method people learn as children. Using an **Insertion Sort**, you arbitrarily select two cards and order them as though they were the only cards being **Sorted** (e.g., "C" and "X"). These two cards begin the **Master Stack**. You introduce a third card (e.g., "P") and fit it within the **Master Stack** in the appropriate spot (e.g., "C," "P," "X"). This system works well for a while but eventually slows down, as you have to **Sort** through an increasingly large **Master Stack** to find the spot where the newest card belongs.

Don't completely disregard the **Insertion Sort**. While not the best choice for **Sorting** large unordered sets, it works well for small ones (ten or fewer items) and for sets that are already mostly in order with just a few outlying items. An example of the latter would be a deck of ordered cards labeled "A" through "Z," but with a second small stack of cards that had fallen out of the main deck (e.g., "L," "N," "B," and "R"); in this case, it makes sense to go through the ordered **Master Stack** and replace the outliers in the appropriate spots as you come across them.

The Bubble Sort

The **Bubble Sort** works by comparing two successive cards, switching them if they're out of order, and moving on to the next pair. Imagine you've picked up off the ground the same randomized twenty-six cards ("A" through "Z") and stacked them; they'd obviously be out of order. The first two cards in the pile are "M" and "A." You switch them so that "A" becomes the first card and "M" becomes the second. You then shift your focus one card deeper into the stack such that "M" becomes the first card of the pair, and "T" becomes your second. These are in the correct order, so no action is necessary. The third pair, "T" and "P," needs to be switched. Note that the second card of each **Sorted** pair always becomes the first card of the next pair.

When you reach the end of the stack, you go back to the beginning and go through it again, performing the same switching action for each pair. This is repeated until you can go through the stack in its entirety without having to perform any swaps. Here lies the inherent inefficiency of this type of **Sort**: it requires multiple passes through the entire stack to complete the task.

For this reason, the **Bubble Sort** will not usually be the best **Manual Shorting** choice, but it can prove useful if you encounter the real-world restriction of having limited physical space to work (for instance, if you're organizing items in a small room where you can't spread them out across the floor).

The Quicksort

The **Quicksort** transcends the linear, single-stack methods and splits the stack into multiple smaller stacks to increase efficiency. Let's revisit the stack of twenty-six cards labeled "A" through "Z."

With a **Quicksort**, you ideally want to end up with two equal stacks. A **Pivot Point** is a rough mid-point where the full set is split. In the "A" through "Z" stack, the **Pivot Point** should be between the letters "M" and "N." As you go through the unsorted stack of cards, any "A" through "M" cards are put in one stack, and any "N" through "Z" cards in another. The end result should be two stacks of equal size. You then repeat this action on each of the two stacks. The "A" through "M" stack is split it into two (roughly) equal stacks using the **Pivot Point** "F," and the same is done to the "N" through "Z" stack using a **Pivot Point** around "S."

This is repeated until you reach a suitable level of granularity. In a computer-guided **Quicksort**, this would mean twenty-six stacks containing one card each (the "A" stack, the "B" stack, etc.) that can then be stacked one on top of the next to arrive at a final, **Sorted** stack. In the real world, you can stop when you have stacks small enough to order quickly (you can most likely **Sort** the cards "A" through "F" in a few seconds), and then pile these **Sorted** stacks atop one other.

The **Quicksort** can be inefficient in terms of space requirements, but it's certainly a useful **Sort** to know. Moreover, the concept of breaking a single stack into multiple smaller stacks will be useful for understanding some of the more human-efficient **Sorting** methods you'll learn about shortly.

The Merge Sort

In a **Merge Sort**, you take the first few items (about six is good) from an unordered set and **Sort** them. This small ordered stack (called a **Preliminary Stack**) is set to one side, and you return to the unordered set to create another **Preliminary Stack**. Once two of these small ordered stacks are created, they're merged into one ordered **Master Stack**. You continue creating new **Preliminary Stacks** and merging them with the **Master Stack** until the **Master Stack** has absorbed all unsorted cards.

On its own, the **Merge Sort** is a bit inefficient, but familiarity with its principles is important for understanding the next type, the **End-Merge Sort**

The End-Merge Sort

The **End-Merge Sort**—a bastardized version of the classic **Merge Sort**—is not a recognized computer science **Sorting** methodology, but is useful in many real world, human-performed **Manual Sorting** situations.

End-Merge Sorts start out the same way as the **Merge Sort**: once the first sorted **Preliminary Stack** is generated, it's placed to the side and another one is started. Unlike the **Merge Sort**, the **Preliminary Stacks** are not merged into a **Master Stack** as you go along; instead, the **Preliminary Stacks** are generated and placed side-by-side. After all the **Preliminary Stacks** are created and no unsorted cards remain, you then merge them into a **Master Stack**.

Again using the "A" through "Z" cards example: after creating all the **Preliminary Stacks**, you have several ordered piles with the lowest card from each (the card closest to "A") on top, face up. Like a game of solitaire, you scan the visible cards, select the lowest one (it should be the "A" card), and extract it from its **Preliminary Stack**. This is the first card in your **Master Stack**. You again look at the top card of each **Preliminary Stack** and extract the lowest (it should be the "B" card) to add to the **Master Stack**. This process is repeated until no cards remain in any of the **Preliminary Stacks** and the result is a **Sorted Master Stack**.

The downside to the **End-Merge Sorts** is that multiple stacks can require a good deal of physical room, which isn't always available. Otherwise, it's an efficient and easy **Manual Sorting** method.

Please note that the applicability of these **Sorts** depends on the idea that the items to be **Sorted** are one-dimensional; in other words, there is only one criterion by which the items are compared. In the card example, this criterion is alphabetical order. These methods could also work for kids in a classroom (ordered by height), lights (by brightness), birds (by wingspan), and so on.

Solving More Complex Manual Sorting Problems

For many real life **Manual Sorting** problems, the **End-Merge Sort** is a good place to start; it's simple enough that you can keep track of what you're doing, but it's still powerful and relatively efficient. This type of **Sort** will be useful as you get into more complex **Manual Sorting** problems.

In a simplified version of the CD example discussed earlier, assume you only have one CD per band (so there's no need to worry about album release order), no compilation disks, and no albums by bands whose names begin with a number. How would you sort this?

It quickly becomes apparent when you come across two artists whose names begin with the same letter that sub-criteria are needed to perform a **Sort**. For instance, "Abba" and "AC/DC" both start with A; however, "Abba" should come before "AC/DC." The **Sort** defers to the next letter, and "B" (from A<u>b</u>ba) comes before "C" (from A<u>C</u>/DC). You might have to dig several letters deep into the names, like with "<u>Lo</u>s <u>Lo</u>bos" and "<u>Los Lon</u>ely Boys."

You could begin by performing a basic **End-Merge Sort**, which results in several smaller stacks of CDs. These smaller stacks would then be merged into one **Master Stack**. However, the sheer number of items being **Sorted** affects the way you should split the initial stack into **Preliminary Stacks**. The smaller the total collection, the fewer **Preliminary Stacks** needed, and the less granular each can be. Forty CDs could produce about seven **Preliminary Stacks,** containing "A to D" CDs, "E to H" CDs, etc. By contrast, three thousand CDs could very well produce individual stacks for each letter: "A," "B," "C," etc.

Solving Highly Complex Manual Sorting Problems

Now let's look at a more complicated version of the CD **Sorting** problem. Again assume there are no compilation disks or artists whose names begin with a number, and that you'll need to **Sort** by artist name. As above, two bands whose names begin with "A" should then be **Sorted** based on the subsequent letters. This time, however, imagine you have more than one album per band, and once the bands are **Sorted**, each bands' albums must be ordered by release date (earliest to latest).

This illustrates an important point. You can get creative and employ several separate **Sorts** based on your needs. With the first (an artist's-name-based **End-Merge Sort**), you can go from a single unsorted stack to several independently-**Sorted Preliminary Stacks**, and then merge them all into an ordered **Master Stack** with no regard for the chronological order of each artist's albums. Next, you can go linearly through the **Sorted Master Stack**, and whenever you happen upon an artist with multiple albums, break their catalogue out of the **Master Stack** into a separate **Temporary Stack**. From here, an **Insertion Sort** can be used to order the **Temporary Stack**. Once you're finished, you can reinsert the now-ordered **Temporary Stack** back into the **Master Stack**.

Use the right tool for the job, and don't be afraid to mix and match.

There are countless ways to **Sort**, but those described here are arguably the most useful for **Manual Sorting** and practical applications.

Quick Review

1. <u>**Sorting**</u>: There are several methods for **Manual Sorting** (sorting real objects without the assistance of third-party tools), such as the **Insertion Sort, Bubble Sort, Quicksort, Merge Sort**, and **End-Merge Sort**. Different situations may call for different methods, based on the complexity and number of items being **Sorted**.

Review and Development: Section 2

In *Section 2*, you learned some practical math skills that can serve you well in personal and professional life. Let's review what you learned and spend some time honing your skills.

Review

1. **Guestimation**: When performing any sort of estimation, it's important to resist the urge to guess or rely on intuition. Force yourself to take measurements and use any available data to make rough calculations. This partially-informed type of guessing is called **Guestimation**.

2. **Probability**: **Probability** is the likelihood that an event will occur, assuming that you don't know for certain whether or not the event will *definitely* occur. **Probability** is measured on a scale of 0 ("not happening") to 1 ("definitely happening").

3. **Practical Probability**: **Practical Probability** is the art of applying **Probability** principles—often somewhat loosely—to everyday situations.

4. **Basic Probability**: When computing basic **Probability**, you are simply comparing the likelihood of an event occurring with the likelihood of other events occurring (given that one such event must occur).

5. **Simple "Or" Events**: When computing a simple "or" event, you must consider the likelihood of *any* (but not necessarily *all*) of two or more events occurring, so you add the occurrences' individual likelihoods together.

6. **"Or" Events with Duplicates**: Some "or" events contain duplicates. In other words, calculating **Probability** as if it were a pure "or" problem would double-count the likelihood of certain outcomes. After adding the individual events' **Probabilities** together (an **"Or" Subtotal**), you must then subtract the duplicate events (called the **Duplicate Set**).

7. **Successive or Simultaneous Events**: **Successive Events** and **Simultaneous Events** are events that are related in that they must all occur to meet your criteria of success. To compute the likelihood of two (or more) **Successive** or **Simultaneous** events occurring, multiply the likelihood of each individual event occurring independently.

8. **The Math Behind Options**: To understand the math behind an option is to better understand its scale, context, and potential impact; it informs your decision-making freedom.

9. **"Determining the Number of Possible Ordering Options" Problems**: To determine the number of possible ways to order a set of objects, tasks, or events, take the total number of items and calculate its "factorial:" multiply the number by "itself minus 1," then by "itself minus 2," and so on, until you reach the number 1. For example, "4 factorial" is equal to $4 \times 3 \times 2 \times 1$, which is 24.

10. **"Determining the Number of Options Given Limitation" Problems**: To determine your possible choices when considering which items (a **Limitation Set**) you'd like to select from a larger set of items, use the formula $C = T!/[(T - L)! \times L!]$. In this formula, C represents your choices, T represents the total number of options, L represents the **Limitation Set**, and $!$ means "factorial."

11. **Quick Percentages**: To translate a fraction into a percentage, first round the fraction, taking into account the level of

precision required by the situation. Simplify the fraction, ideally ending up with rounded one- or two-digit numbers. Identify the relationship between the denominator and 100, and apply that relationship to the numerator. This method is very fast but has the potential to produce somewhat imprecise answers.

12. **Value Comparison**: **Value Comparison** is the oft-required skill of assessing and comparing the relative values of two or more like things. Varying levels of precision and speed can be used depending on the situation. Often, reducing the problem to simpler terms only marginally impacts the answer's precision.

13. **Simple Value Comparison**: **Simple Value Comparison** involves the comparison of two or more things with a shared unit of measurement.

14. **Multi-Faceted Value Comparison**: **Multi-Faceted Value Comparison** refers to problems requiring additional operations or unit conversion before two or more things can be compared properly.

15. **Sorting**: There are several methods for **Manual Sorting** (sorting real objects without the assistance of third-party tools), such as the **Insertion Sort**, **Bubble Sort**, **Quicksort**, **Merge Sort**, and **End-Merge Sort**. Different situations may call for different methods, based on the complexity and number of items being **Sorted**.

Development

Spend six days developing the skills you learned in *Section 2*.

Day 1

On *Day 1*, work on developing your **Guestimation** skills. First, practice on your own; get into the habit of **Guestimating** things in the world around you, from the number of letters or words on a page to the number of pens in a drawer or flowers on the wallpaper. The world is full of things that are composed of other, smaller objects or patterns. **Guestimate** quantity whenever you can, and then take the time to check your **Guestimate** against the real answer.

Day 2

On *Day 2*, focus on **Probability**. Try the following problems (the answers are below). Determine each problem's type, then solve it. These should be easy by now.

1. Johnny and Christian are twin spies who are about to drive (separately) to Virginia to visit their parents for the holidays. Kronolus, the head of an international jewel theft ring, has planted a bomb in one of their cars. Assuming they'll both definitely start their cars and the bomb will reliably detonate when that car is started, what is the probability that one of them will end up a victim of Kronulus' twisted plot?

2. Donnie is a landlord who owns 330 properties. He is planning to evict any tenants who have stopped paying him rent (88 tenants), violated health codes (12 tenants), or are cooking methamphetamines (17 tenants). Note that 7 tenants are *both* violating health codes and failing to pay rent. You live in one of Donnie's properties and suffer from amnesia, so you are unaware of your past activities. You may very well fall into one

of these categories. What is the probability that you will be evicted?

3. Jill begins her first semester at juggling school just as her brother Kyle begins his first semester at unicycle school. They are both fickle, indecisive transients who each have only a 40% chance of graduating, so what is the probability that both will drop out? What is the probability that both will graduate? What is the probability that one will graduate while the other drops out?

4. Lucy is riding in a homemade blimp over Nebraska when a hunter in the woods below drops a large knife. A hawk picks the knife up and flies directly into the blimp, piercing it. Lucy jumps from the falling blimp and happens to land on the back of a hay truck. If the chances of the eagle picking up the knife and piercing the blimp are 1/9,000 and the chances of Lucy landing on the hay truck are 1/7,000, what are the chances of the two events transpiring as they did?

Answers:

1. This is a **Simple "Or" Problem**, as either of the twins' cars exploding is considered a successful event. 1/2 (the chances of Johnny blowing up) + 1/2 (the chances of Christian blowing up) = 2/2 or 1. There is a 100% chance that one of the two will fall victim to the murderous plot.

2. This is an **"Or" with Duplicates** problem. If you add up all the tenants who have done something to get evicted (88 + 12 + 17) and subtract the number of tenants who have been counted twice (7), you are left with 110 tenants being evicted. 110 out of 330 total tenants is 1/3, so you have a 33% chance of being evicted.

3. This is a **Simultaneous Events** problem. If Jill and Kyle each have a 40% (4/10) chance of graduating, it follows that they each have a 60% (6/10) chance of dropping out. To determine the chances of *both* dropping out, you must multiply 6/10 by 6/10, which gives you 36/100. There's a 36% chance of *both* Jill and Kyle dropping out. There's a 16% chance of them *both* graduating (4/10 times 4/10 = 16/100). Note that 36% + 16% does not equal 100%, because there are two other possibilities; Jill may graduate while Kyle drops out, or Kyle may graduate while Jill drops out. The chances of either possibility occurring are 24% (6/10 times 4/10 = 24/100). When added, the chances of all possibilities (none graduate + both graduate + only Jill graduates + only Kyle graduates = 36% + 16% + 24% + 24%) equal 100%, because *one* of those cases definitely has to occur.

4. This is a **Successive Events** problem; the odds of both events occurring are significantly smaller than either of the two events occurring individually. You must multiply 1/9,000 (the chances of the eagle/blimp incident) by 1/7,000 (the chances of the hay truck landing), to arrive at 1/63,000,000. Lesson learned: Lucy is one lucky lass.

After solving these problems, devise one or two problems of each type on your own; this will reinforce your understanding of each problem type's unique characteristics.

Day 3

Day 3 is dedicated to options. Work through the following problems (the answers are below):

1. Vincenzo is a professional competitive eater who is preparing for a championship. He visits a local Chinese buffet with the intention of devising a strategy that will allow him to eat the most food possible without feeling full. He circles the buffet and breaks the foods down into five categories: meats/proteins (M), fatty foods (F), carbohydrates (C), veggies (V), and soups (S). Vincenzo decides to experiment with food order and plans to visit the buffet as many times as necessary to eat foods in every possible order. He sits down with his notes to plan. "Okay," he says, "MFCVS first. Then MFCSV. Then MFVSC. Hmmm." He quickly realizes this may require more buffet trips than he originally thought. How many possible combinations are there?

2. After recovering from triple-bypass surgery, Vincenzo, a former competitive eater, decides to change his lifestyle and adopt healthier habits. He wants to begin slowly, so he makes a list of his ten favorite unhealthy foods and decides to continue eating four of them, banishing the other six. The ten items on the list are: pizza, cheese steaks, barbecue pork sandwiches, ice cream, brownies, root beer, burritos, falafel sandwiches, Oreo cookies and hot chocolate. In narrowing this list down to four items, how many possible combinations can he make?

Answers:

1. This is a "**Determining the Number of Possible Ordering Options**" problem. There are five options, and 5! (five factorial) is 120. Vincenzo should probably reassess his plan.

2. This is a "**Determining the Number of Options Given Limitation**" problem. Using the formula $C = T!/[(T - L)! \times L!]$, this breaks down to $3{,}628{,}800/(720 \times 24)$, which equals 210;

thus, there are about 210 possible four-food combinations Vincenzo can consider appropriate for his new diet.

Again, after solving these problems, devise one or two problems of each type on your own; this will reinforce your understanding of each problem type's unique characteristics.

Day 4

On *Day 4*, develop your ability to quickly translate fractions into percentages. Work through the following problems (the answers are below). Exact answers aren't necessary; just try to get reasonably close. Focus on speed over precision.

1. 23/78
2. 2/908
3. 234/982
4. 38/1,298
5. 23,789/78,941
6. 82,789/98,723

Answers:

1. 29.5%
2. 0.2%
3. 23.8%
4. 2.9%
5. 30.1%
6. 83.8%

Keep working on this skill. Come up with your own fractions and try to translate them to percentages as quickly as possible.

Day 5

Spend *Day 5* developing your **Value Comparison** skills. Work through the following problems (the answers are below). These are a bit more complex and difficult than the ones you've done, but they will force you to think creatively.

1. Cecilia is a tourist visiting Morocco, where she is haggling with a street vendor named Hassan. The vendor is trying to sell her a bootleg copy of Eddie Murphy's 2007 comedy *Norbit* for $1.25 USD. Another vendor, Amal, overhears the conversation and chimes in, telling her that he offers the same film for $1. Much to Cecilia's surprise, the two vendors get into a heated exchange. When the dust settles, she has the option of buying 3 Eddie Murphy DVDs from Hassan for $2.90, or 7 from Amal for $5.05. Assuming Cecilia is more interested in a good bargain than she is the number of DVDs, which vendor should she choose?

2. Angelica wants to hire a private investigator to spy on her cheating husband. She speaks with Aaron, who says he'll take the job at a rate of $40 per hour, plus fuel, tolls, meals, etc. All these extras cost about $30/day. Aaron works five-hour days. Another investigator, Will, also works five-hour days, but charges a flat rate of $60 per hour, with no additional fees. Assuming Angelica wants her husband followed for three days, which investigator is a better bargain?

Answers:

1. This is a **Simple Value Comparison** problem. $2.90/3 vs. $5.05/7 can be simplified to 3/3 and 5/7. To make these numbers directly comparable, the fractions need to be multiplied by a factor that will make the denominators equal approximately 100. The denominators are 3 and 7, and 3 × 33 = 99 and 7 × 14 is 98, so you end up with 99/99 vs. 70/98, or $1 vs. $0.70 per movie. Since the actual, exact answers are $0.97 vs. $0.71, the rough answers offer a reasonable basis for comparison. Amal is offering a better deal.

2. This is a **Multi-Faceted Value Comparison** problem with additional operations. Aaron's charge for three 5-hour days at $40 per hour would equal $600. Will's invoice for the same time would be $900. However, Aaron's daily "extras" charge would add an additional $90 to the bill ($30 × 3 days), bringing his total to $690. $690 for 3 days vs. $900 for 3 days; even with the charges, Aaron offers a better bargain.

Day 6

Spend *Day 6* reviewing the different **Sorting** methods. Make sure that you understand each and come up with at least one scenario that would be a good fit for each method. In addition, purposely jumble a group of ordered objects, such as books, records, or recipes, and put them back in order using each of the **Sorting** methods discussed.

Before moving on, go back and review any parts of this section that were particularly confusing or difficult for you.

Section 3: New Approaches to Multiplication

In this section, you'll venture into a strange world of alternative multiplication methods that are probably unknown to you.

You'll begin by learning some shortcuts for handling common types of multiplication problems, and then you'll explore Jakow Trachtenberg's ingenious methods for multiplying any number by the numbers five through twelve. The need to multiply by such small numbers is fairly common (more common than the need to multiply two larger numbers), so these methods provide practical value.

Though some of this material may seem intimidating, don't worry—it's not rocket science (*this* is rocket science: $F = lm[dot] \, v_e + [p_e - p_a] \, A_e$). With dedicated practice and the right approach, you'll find that the types of calculation taught here aren't so tough after all.

You'll end the section by formulating a sure-fire plan for quickly selecting the best multiplication method for any given practical problem type.

I know it's been a while since algebra class, so be aware that we will be using variables in this section. When talking about multiplication by specific numbers, you may come across the variable "*n*"; "*n*" is simply a placeholder that refers to a value you intend to multiply by another number. For example, you know that multiplying any number by twenty is the same as multiplying that same number by two and then by ten. Using algebraic notation, this could be expressed as *20n = 10(2n)*.

This is the extent to which you'll need algebra in this book. If you still think you need a refresher, perform an Internet search for "introduction to algebra." There is plenty of information out there to get you caught up.

Let's begin.

Lesson 1: Multiplication Shortcuts (Part 1)

Sometimes blunt advice from an absurdist post-apocalyptic children's cartoon can be surprisingly poignant.

Let's learn some practical **Shortcuts** for multiplying certain types of small, common numbers. Those featured were selected due to their applicability and simplicity. They've been given memorable names, because it never hurts to make something more vivid or "catchy."

You'll certainly suck at these before you become good at them.

The Teeny-Teeny

Numbers in the "teens" (11 through 19) are common in daily life. Luckily, there's a trick that will allow you to multiply two teen numbers in just a few seconds.

The process:

1. Take the ones place (rightmost digit) from the smaller number and add it to the larger number.
2. Add a 0 to the end of this sum.
3. Hold this combined number in your mind.
4. Multiply the ones place (rightmost) digits of the two original numbers.
5. Add the product of step 4 to the number from step 3.

That can seem daunting at first, so let's go through the steps again with an example in mind: 15 × 13. Both numbers are "teens," so they qualify for this method.

1. Take the one's place (rightmost digit) from the smaller number (the 3 from the 1<u>3</u>), and add it to the larger number (3 + 15 = 18).
2. Add a 0 to the end of this sum: 180.
3. Hold the 180 in your mind.
4. Take the rightmost (ones place) digits of the two original numbers (1<u>5</u> and 1<u>3</u>), and multiply them (5 × 3 = 15).
5. Add the outcome of step 4 (15) to the number you're holding in your head (180). 180 + 15 = 195.

That's not so bad, is it? In fact, you can do this in your head, with no tools or pencil and paper. Try to multiply 12 × 19 using only your head.

1. 2 + 19 = 21
2. Add the 0: 210
3. Hang on to the 210
4. 2 × 9 = 18
5. 210 + 18 = 228

Answer: 228.

Congratulations! This is your maiden voyage into the world of **Mental Calculation**. With practice, multiplying using the **Teeny-Teeny** method will become a quick and painless process.

Perform the following calculations; the answers are below. Don't cheat yourself; figure out and write down the answers, then check them. Only write down the answer digits, never the problem steps or work. If you've made a mistake, try to figure out where you went wrong.

1. 15×15
2. 16×13
3. 12×19
4. 18×17
5. 19×11

Answers:

1. 225
2. 208
3. 228
4. 306
5. 209

The Splitsy-Doubly

If you need to multiply two numbers and at least one of them is a power of two (2, 4, 8, 16, 32, 64, 128, 256, 512...), you can work through the problem by repeatedly halving the "power of two" number while doubling the other. Repeat this "split" and "double" step until the problem is simple enough to solve. To perform this **Shortcut** properly, you need to feel comfortable **Splitting** numbers in half (that is, dividing them by two); this skill was covered in *Fundamental Mathematical Concepts*. Let's look at an example: 16×23.

16 is a power of 2, so you're going to repeatedly split this while doubling its counterpart.

1. $(16/2 = 8) \times (23 \times 2 = 46)$
2. $(8/2 = 4) \times (46 \times 2 = 92)$
3. $(4/2 = 2) \times (92 \times 2 = 184)$
4. $2 \times 184 = 368$

Answer: 368.

With practice, you'll learn to recognize powers of two instantly and can then decide whether this method would be the best choice given the circumstances. Eventually, when you recognize a power of two, you'll also be able to determine *which* power of two it is and double the other number that many times. For example, when faced with 16×23, you will be able to say, "Oh, 16; that's a power of 2. In fact, 16 is 2 to the 4th, so I have to multiply 23 by 2 exactly 4 times." This allows you to skip the "splitting" aspect and concentrate on "doubling."

When starting out, it can be useful to use your fingers to keep track of which power of two you're working with; with the example of 16, you'd begin with four fingers extended, then three, then two, etc. Let's try a more difficult example: 64×31.

1. $(64/2 = 32) \times (31 \times 2 = 62)$
2. $(32/2 = 16) \times (62 \times 2 = 124)$

3. $(16/2 = 8) \times (124 \times 2 = 248)$
4. $(8/2 = 4) \times (248 \times 2 = 496)$
5. $(4/2 = 2) \times (496 \times 2 = 992)$
6. $2 \times 992 = 1{,}984$

Answer: 1,984.

Technically, this method works when the problem contains any even number; it's not necessarily reserved for powers of two. However, you can end up running into issues with non-"power-of-two" problems, as you may have to split an odd number along the way. In these cases, the **Splitsy-Doubly** can still be used to turn one of the two numbers into a smaller, more manageable number, transforming a daunting problem into an easier one. Let's look at an example: 14×22.

$(14/2 = 7) \times (22 \times 2 = 44)$

You're stuck with an odd number, but at the same time, the new problem, 7×44, is now arguably easier; and when you soon learn how to solve this problem using Trachtenberg's methods for speed multiplication, it will be even easier.

Answer: 308.

This only works some of the time. Let's next look at a non-"power of two" example where using this method will *not* help: 38×21.

$(38/2 = 19) \times (21 \times 2 = 42)$

You now have to multiply 19 by 42. That's still ugly, and really no simpler than where you started off (38×21). Remain mindful that the **Splitsy-Doubly** is an available option, but it's only suggested as your primary method for any multiplication problems involving a power of two.

Perform the following calculations; the answers are below. Don't cheat yourself; figure out and write down the answers, then check them. Only write down the answer digits, never the problem steps or work. If you've made a mistake, try to figure out where you went wrong.

1. 16×3
2. 64×9
3. 32×12
4. 54×16
5. 128×22
6. 256×14

Answers:

1. 48
2. 576
3. 384
4. 864
5. 2,816
6. 3,584

The Double-Double

To multiply any number by four, perform this simple two-step operation:

1. Double the number, and

2. Double it again.

This would be expressed algebraically as *4n = 2(2n)*.

It may seem obvious, but it's easy to forget that you can think of "4" as "2 × 2". This method is best suited for *n* × *4* problems where *n* is small enough to multiply by 2 somewhat easily.

Here's an example of an optimal **Double-Double** method use case: 55 × 4.

1.	55 × 2 = 110
2.	110 × 2 = 220

Answer: 220.

See? For most, multiplying 55 by 4 is considerably more intimidating than either of the two individual steps (55 × 2 and 110 × 2). Let's look at another example: 74 × 4.

1.	74 × 2 = 148
2.	148 × 2 = 296

Answer: 296.

If precision isn't required, you can round before multiplying, turning 74 into 75, leveraging your familiarity with quarters and currency.

1.	75 × 2 = 150
2.	150 × 2 = 300

Generally, the **Double-Double** method is most useful for multiplying 4 by numbers smaller than 999 (or **Familiar Numbers** of any size). An example of a large-but-**Familiar** number would be 2,800 (2,800 × 2 = 5,600; 5,600 × 2 = 11,200). An example of a less simple larger number would be 1,731. When numbers are too large or complicated for this method, you can round them if appropriate, or use the **Trachtenberg System** long multiplication method, which you'll learn shortly.

Perform the following calculations; the answers are below. Don't cheat yourself; figure out and write down the answers, then check them. Only write down the answer digits, never the problem steps or work. If you've made a mistake, try to figure out where you went wrong.

1. 71 × 4
2. 19 × 4
3. 114 × 4
4. 53 × 4
5. 516 × 4
6. 890 × 4
7. 2,100 × 4

Answers:

1. 284
2. 76
3. 456
4. 212
5. 2,064
6. 3,560
7. 8,400

The Five-Ten Split

To multiply any number by five, perform this simple two-step operation:

1. Cut the number in half, and
2. Multiply by ten.

Depending on how your mind works, it may be easier to think of the second step as "moving the decimal one place to the right" instead of multiplying by ten. Either way, it's the same principle. This would be expressed algebraically as $5n = 10(\frac{1}{2}n)$.

Remember how I stressed the importance of learning to **Split** numbers quickly a little while ago? The **Five-Ten Split** is a prime example of that skill's usefulness.

Let's try an example: 84×5.

1. $84/2 = 42$
2. $42 \times 10 = 420$

Answer: 420.

Let's look at an odd number problem: 75×5.

1. $75/2 = 37.5$
2. $37.5 \times 10 = 375$

Answer: 375.

As you can see, this method handles odd number problems (especially those involving small numbers) somewhat gracefully despite the involvement of the decimal place; this introduces only a marginal increase in complexity.

Generally, the **Five-Ten Split** method is useful for multiplying 5 by any number smaller than 9,999. Numbers that split more gracefully (those ending in 0, for instance) tend to cooperate better. For example, $3,020 \times 5$ is arguably simpler to solve than $3,196 \times 5$. Let's break them down.

$3,020 \times 5$:

1.	$3,020/2 = 1,510$
2.	$1,510 \times 10 = 15,100$

$3,196 \times 5$:

1.	$3,196/2 = 1,598$
2.	$1,598 \times 10 = 15,980$

Which was easier for you?

When a problem involves numbers that are harder to **Split** (such as 3,393), you can round or, for precise answers, rely on the **Trachtenberg System** method for multiplication by 5, which you'll learn soon.

Perform the following calculations; the answers are below. Don't cheat yourself; figure out and write down the answers, then check them. Only write down the answer digits, never the problem steps or work. If you've made a mistake, try to figure out where you went wrong.

1. 26 × 5
2. 124 × 5
3. 72 × 5
4. 55 × 5
5. 388 × 5
6. 1,540 × 5

Answers:

1. 130
2. 620
3. 360
4. 275
5. 1,940
6. 7,700

Next, you're going to learn some more **Shortcuts**; first, take some time and get used to the ones covered in this lesson. Move on when you're comfortable.

Quick Review

1. **Multiplication Shortcuts**: For some types of practical multiplication problems, a series of situation-specific **Shortcuts** will apply. These include **Teeny-Teeny, Splitsy-Doubly, Double-Double**, and the **Five-Ten Split**.

Lesson 2: Multiplication Shortcuts (Part 2)

In this lesson, you'll learn some more **Shortcuts** for solving specific types of common multiplication problems. Some may prove to be a bit more challenging than those in the last lesson.

Nine-to-Ten and Down Again

To multiply any number by nine, you can perform a simple two-step operation:

1. Multiply it by ten (in other words, place a zero at the right side of the number), then
2. Subtract the original number from the product of step 1.

This would be expressed algebraically as $9n = 10n - n$.

To put this in context, consider the problem 9×4. You know $4 \times 10 = 40$ (step 1), and 40 minus 4 (step 2) is 36. While this may seem obvious with such small numbers, the method scales wonderfully. Let's take a look at another simple example: 30×9.

1. $30 \times 10 = 300$
2. $300 - 30 = 270$

Answer: 270.

That's all there is to it. Let's look at a more complex example (in that the second step crosses over a "hundreds" line): 44×9.

1. $44 \times 10 = 440$
2. $440 - 44 = 396$

Answer: 396.

Perform the following calculations; the answers are below. Don't cheat yourself; figure out and write down the answers, then check them. Only write down the answer digits, never the problem steps or work. If you've made a mistake, try to figure out where you went wrong.

1. 16×9
2. 22×9
3. 90×9
4. 54×9
5. 112×9

Answers:

1.   144

161

2. 198
3. 810
4. 486
5. 1,008

The Quarter Pounder

The number 25, especially in reference to 25% or "one quarter" of a whole, appears all around us. This quarter's earnings. The third quarter of the big game. A pizza shared among four friends. As a result, you often have to perform mathematical procedures involving the number 25. Luckily, there's a shortcut for that.

To multiply any number by 25, perform this simple two-step operation:

1. Multiply it by 100, then
2. Divide the product of step 1 by 4.

Expressed algebraically, this works out to $25n = 100n/4$. Let's try an example: 20 × 25.

1. $20 \times 100 = 2000$
2. $2000/4 = 500$

Answer: 500.

Easy enough, right? With more complex problems, the division step can be daunting. Let's look at the following example: 17 × 25.

1. $17 \times 100 = 1,700$

Answer: 425.

If that final step (1,700/4) was challenging to do quickly, keep in mind that you can divide 1,700 by 2 (850), and then divide it by 2 again (425); this is sort of a reversed version of the **Double-Double** shortcut. As you become more comfortable with numbers and their relationships, you can exercise a higher degree of creativity.

Let's do another, using this **Double-Double** trick: 620 × 25.

> 1. 620 × 100 = 62,000
> 2. 62,000/4 = (62,000/2 = 31,000 and 31,000/2) = 15,500

Answer: 15,500.

Perform the following calculations; the answers are below. Don't cheat yourself; figure out and write down the answers, then check them. Only write down the answer digits, never the problem steps or work. If you've made a mistake, try to figure out where you went wrong.

1. 12 × 25
2. 21 × 25
3. 43 × 25
4. 122 × 25
5. 666 × 25

Answers:

1. 300
2. 525

3. 1,075
4. 3,050
5. 16,650

The Jigsaw

This is a cool **Shortcut**. It's called the **Jigsaw** because two parts must fit together perfectly in order for it to work. This one is a little complex, but with only a little practice, it becomes much easier.

If two two-digit numbers share the same tens place digit (like how <u>3</u>1 and <u>3</u>9 share 3), and the two ones place digits add up to 10 (like how the ones places of 3<u>1</u> and 3<u>9</u> add up to 10), the **Jigsaw** method can be used to multiply them quickly. There are three steps involved. Continuing with the 31 × 39 example:

1. First, multiply the shared tens place digit by "itself plus one." Since the two leftmost digits are both 3's, then you multiply 3 × 4.
2. Next, multiply the two ones place digits. Here, this is 9 × 1 (3<u>1</u> × 3<u>9</u>). When this step results in a single-digit product, include a preceding 0 (so 9 becomes <u>0</u>9).
3. Finally, place the product of the first step (3 × 4 = 12) to the left of the product of the second step (9 × 1 = 09). The final answer is 1,209.

Here are the three steps without commentary:

1. 3 × (3+1) = 12
2. 1 × 9 = 09

Answer: 1,209.

Let's do another example: 45 × 45.

1. 4 × (4 + 1) = 20
2. 5 × 5 = 25
3. Place 20 in front of 25.

Answer: 2,025.

Let's do one more: 93 × 97.

1. 9 × (9 + 1) = 90
2. 3 × 7 = 21
3. Place 90 in front of 21.

Answer: 9,021.

Perform the following calculations; the answers are below. Don't cheat yourself; figure out and write down the answers, then check them. Only write down the answer digits, never the problem steps or work. If you've made a mistake, try to figure out where you went wrong.

1. 13 × 17
2. 24 × 26
3. 37 × 33
4. 89 × 81
5. 75 × 75

Answers:

1. 221
2. 624
3. 1,221
4. 7,209
5. 5,625

Double Down

There is a simple two-step operation for multiplying any number by twenty:

1. Double the number, and
2. Multiply the product of step 1 by 10.

Again, it seems obvious, but many people attempt this in much more difficult ways; some even imagine the problem written out in traditional "stacked" notation, like you learned in school.

The **Double Down** method is much simpler. Let's try 41 × 20.

1. $41 \times 2 = 82$
2. $82 \times 10 = 820$

Answer: 820

Expressed algebraically, this is *20n = 10(2n)*.

Let's do another: 88 × 20.

1. $88 \times 2 = 176$
2. $176 \times 10 = 1,760$

Answer: 1,760.

There's not much more to say about this **Shortcut** except that the better you are at **Doubling**, the easier this will be.

Perform the following calculations; the answers are below. Don't cheat yourself; figure out and write down the answers, then check them. Only write down the answer digits, never the problem steps or work. If you've made a mistake, try to figure out where you went wrong.

1. 18×20
2. 46×20
3. 630×20
4. 777×20
5. 930×20

Answers:

1. 360
2. 920
3. 12,600
4. 15,540
5. 18,600

FOIL

To multiply any two-digit number by any other two-digit number, you can follow these five steps:

1. Multiply the first digit of the first number by the first digit of the second number and multiply the product by 100.

167

2. Multiply the first digit of the first number by the second digit of the second number and multiply the product by ten.
3. Multiply the second digit of the first number by the first digit of the second number and multiply the product by ten.
4. Multiply the second digit of the first number by the second digit of the second number.
5. Add the products of steps 1 through 4 together.

It sounds like a lot of work, but it's easy if you remember the mnemonic **FOIL**, which stands for "**F**irst two, **O**uter two, **I**nner two, **L**ast two." Let's look at an example: 63×23.

---

1. First two: $\underline{6}3 \times \underline{2}3$
2. Outer two: $\underline{6}3 \times 2\underline{3}$
3. Inner two: $6\underline{3} \times \underline{2}3$
4. Last two: $6\underline{3} \times 2\underline{3}$

Do you see how the steps got their names? Let's work through this example together:

---

1. First two: $\underline{6}3 \times \underline{2}3 = 12$, and $12 \times 100 = 1{,}200$
2. Outer two: $\underline{6}3 \times 2\underline{3} = 18$, and $18 \times 10 = 180$
3. Inner two: $6\underline{3} \times \underline{2}3 = 6$, and $6 \times 10 = 60$
4. Last two: $6\underline{3} \times 2\underline{3} = 9$
5. $1{,}200 + 180 + 60 + 9 = 1{,}449$

It's easier to add these products as you go, turning five steps into four:

---

1. First two: $\underline{6}3 \times \underline{2}3 = 12$, and $12 \times 100 = 1{,}200$. Running total: 1,200

Let's do another: 89×14.

Answer: 1,246.

Though somewhat complex, this **Shortcut** may be the most useful one you'll learn, since it doesn't require special circumstances in order to work.

Perform the following calculations; the answers are below. Don't cheat yourself; figure out and write down the answers, then check them. Only write down the answer digits, never the problem steps or work. If you've made a mistake, try to figure out where you went wrong.

1. 16×22
2. 23×81
3. 67×19
4. 55×82
5. 79×35

Answers:

1. 352
2. 1,863
3. 1,273
4. 4,510
5. 2,765

Quick Review

1. **Multiplication Shortcuts**: There are several situation-specific **Shortcuts** for solving practical multiplication problems. These include **Nine to Ten and Down Again**, the **Quarter Pounder**, the **Jigsaw**, the **Double Down**, and **FOIL**.

Lesson 3: Trachtenberg Multiplication—Multiplying by 11, 12, 6, and 7

Most of us were taught that all multiplication problems can and should be solved using the standard classical method; however, this may not be the fastest or easiest route to a solution.

Let's do some unlearning.

Multiplication is woven into the fabric of life in subtle but consistent ways. For many, multiplication most often comes to us in the form of a single-digit number multiplied by another single-digit number ("Dave has been consistently losing two pounds per week for seven weeks"). The next most common type of practical problem is arguably a single-digit number multiplied by a two-digit number ("The %$#@% bank has been charging me a $20 overdraft fee each day for the last six days?"). Following this line of logic, the next most common type of problem involves a single-digit number multiplied by a three-digit number ("If Sasha reduces her monthly expenditure by $410 per month for nine months, she'll have enough money for a down-payment on that car").

With this in mind, we'll begin with methods for multiplication by common single-digit and low double-digit numbers. Many methods have been explored for solving these types of multiplication problems. Both classic and modern works of **Mental Calculation** literature tend to promote the practice of multiplying from left to right. With a problem such as 3 × 281, the first step of this method is to multiply the single-digit number (3) by the larger number's leftmost digit, followed by the appropriate number of zeros (3 × 200 = 600). The product of that calculation (600) is added to the product of the single digit number and the next digit, again followed by the appropriate number of zeros (3 × 80 = 240). 600 + 240 = 840. The single-digit-number (3) is then multiplied by the rightmost digit of the larger number (3 × 1 = 3), building the final answer (600 + 240 + 3 = 843).

This is fast, simple, and perfectly valid; however, the method becomes increasingly complex when scaled. Multiplying 7 by 135 may be a seven-second job, but multiplying 7 by 135,654 is not; the complexity inherent in holding onto partial answers while performing calculations can draw your attention away from the aggregate final answer, which must be kept in your functional memory as you work.

The **Trachtenberg System** is often more appropriate for this type of problem. Multiplying 7 by 13, 135, or 135,654 all require a single method for finding the product without ever actually multiplying by 7.

You'll begin by multiplying by one- and two-digit numbers that often come up in daily life. The lesson builds slowly from simplistic examples to more difficult ones as you get used to the **Trachtenberg System's** recurring themes. Use a pencil and paper to record the digits of your answers as you figure them out, but *do not write down the intermediate steps.*

Trachtenberg's approach to math looks different than what you may be used to. It bears little resemblance to commonly taught classical math systems, but the challenge bears great rewards. **Trachtenberg System** methods—even the more challenging ones—are far easier to perform in your head than classical math. The graceful handling of remainders alone makes it worthwhile; there are no numbers to carry aside from the occasional one or two. In classical math, multiplying by two-digit numbers produces two rows of answers that need to be added; the **Trachtenberg System** produces solutions in a linear manner and on a single line. This simplicity will prove valuable when you move calculation off-paper and into your skull.

Starting Easy: Multiplying by 11

The **Trachtenberg System** was first popularized in an extensive 1960 work by Ann Cutler and Rudolph McShane called *The Trachtenberg Speed System of Basic Mathematics*. As in the book, you will first become familiar with the **Trachtenberg System** by learning to multiply numbers by 11.

First, imagine a zero before and after the number. Take this example:

$3,231 \times 11$

This problem becomes:

03,23<u>10</u> × 11

The zeroes will be a recurring theme in many of the Trachtenberg multiplication methods you'll learn. You won't actually write them down, but you'll get used to imagining them.

Ignore the 11 entirely, focusing on the other number. Beginning with the rightmost digit (that is, the *actual* rightmost digit, not the added zero), *add each digit to its "neighbor"* (the digit to its right).

03,23<u>10</u>: 1 + 0 = 1
03,2<u>31</u>0: 3 + 1 = 4
03,<u>23</u>10: 2 + 3 = 5
0<u>3,2</u>310: 3 + 2 = 5
<u>03</u>,2310: 0 + 3 = 3

Each added pair (number + neighbor) produces an individual digit of the answer (1, 4, 5, etc.); write these down in order from right to left.

Answer: 3,231 × 11 = 35,541.

"Wait. That's really all there is to it?" Yes. Really, that's it. Let's do another example together: 4,154 × 11.

04,15<u>40</u>: 4 + 0 = 4
04,1<u>54</u>0: 5 + 4 = 9
04,<u>15</u>40: 1 + 5 = 6
0<u>4,1</u>540: 4 + 1 = 5
<u>04</u>,1540: 0 + 4 = 4

Answer: 4,154 × 11 = 45,694. No sweat. Try one yourself: 12,512 × 11.

Did you get 137,632? If not, go back and do it again; work through it slowly if you must.

The only complication is caused by pairs of numbers that—when added—exceed 9. You need to handle the carried number, just like in classic arithmetic. Unlike classic arithmetic, however, you never have to carry large numbers—just the occasional 1 or 2. Especially when starting out, it can be useful to "hold onto" carried numbers using the fingers of your non-writing hand. Let's look at an example where carrying is required: 2,481 × 11.

02,48**10**: 1 + 0 = 1
02,4**81**0: 8 + 1 = 9
02,**48**10: 3 + 8 = 12 (write the 2 down and carry the 1) = 2
0**2,4**810: 2 + 4 = 6 (plus the carried 1 from above) = 7
02,4810: 0 + 2 = 2

Answer: 2,481 × 11 = 27,291.

Did that make sense? Let's look at an example with even more extreme carrying: 297,361 × 11.

0297,36**10**: 1 + 0 = 1
0297,3**61**0: 6 + 1 = 7
0297,**36**10: 3 + 6 = 9
029**7,3**610: 7 + 3 = 10 (write the 0 down and carry the 1) = 0
02**97**,3610: 9 + 7 = 16 (plus the carried 1 from above = 17. Write the 7 down and carry the 1) = 7
0**29**7,3610: 2 + 9 = 11 (plus the carried 1 from above = 12. Write the 2 down and carry the 1) = 2
0297,3610: 0 + 2 = 2 (plus the carried 1 from above) = 3

Answer: 297,361 × 11 = 3,270,971.

Perform the following calculations; the answers are below. Don't cheat yourself; figure out and write down the answers, then check them. Only write down the answer digits, never the problem steps or work. If you've made a mistake, try to figure out where you went wrong.

1.   1,432 × 11
2.   542,314 × 11
3.   938,920 × 11
4.   19,821,895 × 11

Answers:

1.   15,752
2.   5,965,454
3.   10,328,120
4.   218,040,845

As you develop and get more comfortable with this skill, you'll start to abbreviate the steps. When beginning the first problem above (1,432 × 11), you may have thought, "Two plus zero equals *two*. Three plus two equals *five*." If you keep working at it, you'll soon find yourself simply saying, "*two, five...*" The addition becomes an unconscious part of the calculation; you'll see the two digits and automatically add them together to arrive at the individual answer digits.

You may also begin to notice some shortcuts. For instance, the rightmost digit (not counting the imaginary 0s) will always be the first answer digit—it has no neighbor to its right, so it's always being added to 0 (e.g., 1,43<u>2</u> × 11 = 15,75<u>2</u>).

When multiplying any two-digit number by 11, you'll notice another shortcut: if the number's two digits add up to 9 or less, the answer to the problem will be this sum shimmed in between the two digits themselves. For example, 71 × 11 can be done this way because 71 is a two-digit number, and 7 + 1 is less than 9. Simply place the sum of the two digits (8) in between the 7 and the 1, and you arrive at the answer: 7<u>8</u>1. In the same way, 63 × 11 = 6<u>9</u>3 and 52 × 11 = 5<u>7</u>2; 86 × 11 will not work, because the sum of 8 + 6 is greater than 9.

This is an easy method for multiplication by eleven. Even after such a brief introduction, you should have a decent grasp on it. Next you'll learn how to multiply by twelve, which is only marginally more complex.

Multiplying by 12

To multiply a number by twelve, begin at the right, *double each digit*, and *add the neighbor*.

Let's try this with 3,231 × 12. First, surround the multiplier with 0's, just as you did for eleven:

3,231 × 12

...becomes...

<u>0</u>3,231<u>0</u> × 12

Then take each digit, *double it*, and *add the neighbor*:

03,23<u>10</u>: 1 (x 2) + 0 = 2

177

03,23**10**: 3 (x 2) + 1 = 7
03,2**3**10: 2 (x 2) + 3 = 7
03,**2**310: 3 (x 2) + 2 = 8
03,2310: 0 (x 2) + 3 = 3

As with 11, assemble your answer digits from right to left.

Answer: 3,231 × 12 = 38,772.

Try a problem out for yourself: 14,112 × 12.

Did you get 169,344? If not, go back and do it again. Go slowly and be sure to adhere strictly to the method.

Carrying is addressed in the same way as before; simply add the leftover number to the next step. Dive right into a somewhat difficult example: 2,796 × 12.

02,79**60**: 6 (x 2) + 0 = 12 (write the 2, carry the 1) = 2
02,7**9**60: 9 (x 2) + 6 + (the carried) 1 = 25 (write 5, carry the 2) = 5
02,7**9**60: 7 (x 2) + 9 + (the carried) 2 = 5 (carry the 2)
02,**7**960: 2 (x 2) + 7 + (the carried) 2 = 3 (carry the 1)
02,7960: 0 (x 2) + 2 + (the carried) 1 = 3

Answer: 2,796 × 12 = 33,552.

Perform the following calculations; you'll find the correct answers below. Don't cheat yourself; figure out and write down the answers, then check them. Only write down the answer digits, never the problem steps or work. If you've made a mistake, try to figure out where you went wrong.

1.   2,412 × 12

2. $442,614 \times 12$
3. $918,127 \times 12$
4. $69,771,815 \times 12$

Answers:

1. 28,944
2. 5,311,368
3. 11,017,524
4. 837,261,780

Once you're able to solve problems like this quickly, a small, practical calculation such as 12×133 will be no trouble.

Now that you know how to multiply numbers by 11 and 12 using the **Trachtenberg System**, practice it on the numbers you find around you in everyday life. Grab a pencil and paper and multiply any numbers you see by 11 and 12 as quickly as you can. Use the UPC numbers from products in your home and the birthdays of family members. Use the current time, or even the digits that come up after smacking your hand down on a calculator (you can then use that calculator to check your answers).

The world is one giant math problem waiting to be solved; use your imagination.

Even if you despise doing math—like I used to—forcing yourself into this habit will make it something you do without thinking. When you become significantly better at implementing the methods taught in this book, you might actually begin to enjoy doing math.

It's easy to assume that performing math quickly in your head is difficult because so few are able to do it. But if you think about it, no one is born able to type on a QWERTY keyboard, and yet many people can do so very quickly. The only difference between those who lack this skill and those who have mastered it is *time spent practicing correctly*.

Fortunately, practicing calculations on numbers that appear in everyday life reduces the need to dedicate isolated time to this. You can practice during the vast amount of "down time" that happens while doing other things or waiting around. Hone your math skills during boring work meetings, long bus rides, or jogs.

If you're so inclined, feel free to stop here and take a day or two to work on multiplying by eleven and twelve using the **Trachtenberg System**.

Multiplying by 6

If you feel comfortable multiplying by eleven and twelve, you can move on. To multiply a number by six using the **Trachtenberg System**, *add five to each digit if it is odd* (simply keeping the original digit if it's even), and *add half the neighbor*, starting from the right.

That sounds a bit intimidating. First, let's look at an easy example with no odd digits: 22×6.

22×6

...becomes...

$$0220 \times 6$$

...then...

0220: 2 + (1/2 of 0) = 2
0220: 2 + (1/2 of 2) = 3
0220: 0 + (1/2 of 2) = 1
Answer: 22 × 6 = 132

You may think "halving" the neighbor has the potential to cause problems, as odd digits don't split evenly. The solution couldn't be simpler; in these cases, you simply drop the fraction. 2½ becomes 2, 4½ becomes 4, ½ becomes 0, and so on.

With that cleared up, the following example contains a digit that doesn't split evenly. Incidentally, this also means there will be an odd digit, invoking the fancy "add 5" rule described above. Let's try 141 × 6.

$$141 \times 6$$

...becomes...

$$01410 \times 6$$

...then...

01410: (1 + 5, because 1 is odd) + (1/2 of 0) = 6
01410: 4 + (1/2 of 1) = 4
01410: (1 + 5, because 1 is odd) + (1/2 of 4) = 8
01410: 0 + (1/2 of 1) = 0

Answer: 141 × 6 = 846.

Carrying is handled in these problems the same way it was for eleven and twelve; add the previous step's carry to the current step's answer. Let's try a problem that showcases this phenomenon: 2,425 × 6.

2,425 × 6

...becomes...

0<u>2</u>,425<u>0</u> × 6

...then...

0<u>2,425<u>0</u>: (5 + 5) + (1/2 of 0) = 0 (carry the 1)
02,4<u>25</u>0: 2 + (1/2 of 5) + (the carried) 1 = 5
02,<u>42</u>50: 4 + (1/2 of 2) = 5
0<u>2,4</u>250: 2 + (1/2 of 4) = 4
<u>02</u>,4250: 0 + (1/2 of 2) = 1

Answer: 2,425 × 6 = 14,550.

The next example is a bit tougher, as it contains several odd digits: 15,958 × 6.

15,958 × 6

...becomes...

0<u>1</u>5,958<u>0</u> × 6

...then...

015,95<u>80</u>: 8 + (1/2 of 0) = 8
015,9<u>58</u>0: (5 + 5) + (1/2 of 8) = 4 (carry the 1)

015,**9**580: (9 + 5) + (1/2 of 5) + (the carried) 1 = 7 (carry the 1)
01**5,9**580: (5 + 5) + (1/2 of 9) + (the carried) 1 = 5 (carry the 1)
0**1**5,9580: (1 + 5) + (1/2 of 5) + (the carried) 1 = 9
015,9580: 0 + (1/2 of 1) = 0

Answer: 15,958 × 6 = 95,748.

Perform the following calculations; the answers are below. Don't cheat yourself; figure out and write down the answers, then check them. Only write down the answer digits, never the problem steps or work. If you've made a mistake, try to figure out where you went wrong.

1. 3,042 × 6
2. 641,114 × 6
3. 928,247 × 6
4. 59,972,225 × 6

Answers:

1. 18,252
2. 3,846,684
3. 5,569,482
4. 359,833,350

Here's a quick tip: When solving $n \times 6$ problems where n begins with a 1, the final (leftmost) digit of the answer will always be a 0, unless there's carrying involved; this is because [the left-side added 0] + [1/2 of 1] = 0. This tiny shortcut can help speed up the last step of such calculations.

Multiplying by 7

To multiply by seven, *double each digit, add five if it is odd*, and *add half the neighbor*. Again, fractions are dropped from any digits that fail to split cleanly, and carrying is handled the same way as before. Note how the method for multiplying numbers by 7 involves now-familiar concepts from 11, 12, and 6, just in a different arrangement. Nothing here is too complex. Mastery simply requires practice.

Let's do an example problem: 1,921 × 7.

1,921 × 7

...becomes...

<u>0</u>1,921<u>0</u> × 7

Take a deep breath...

01,92<u>10</u>: 1 (x 2) + 5 + (1/2 of 0) = 7

01,9<u>21</u>0: 2 (x 2) + (1/2 of 1) = 4

01,<u>92</u>10: 9 (x 2) + 5 + (1/2 of 2) = 4 (carry the 2)

0<u>1,9</u>210: 1 (x 2) + 5 + (1/2 of 9) + (the carried) 2 = 3 (carry the 1)

<u>01</u>,9210: 0 (x 2) + (1/2 of 1) + (the carried) 1 = 1

Answer: 1,921 × 7 = 13,447.

This can seem complicated at first, but you'll get the hang of it in no time.

Perform the following calculations; the answers are below. Don't cheat yourself; figure out and write down the answers, then check them.

Only write down the answer digits, never the problem steps or work. If you've made a mistake, try to figure out where you went wrong.

1. $4{,}152 \times 7$
2. $711{,}123 \times 7$
3. $923{,}291 \times 7$
4. $29{,}922{,}471 \times 7$

Answers:

1. 29,064
2. 4,977,861
3. 6,463,037
4. 209,457,297

It's time to practice. Make good use of the world around you. Weave practice into the fabric of your life. Constantly and habitually multiply numbers you see by eleven, twelve, and especially your new friends six and seven, until you become reasonably comfortable with them.

Review the methods covered in this lesson as needed over the next few days, practice as much as possible, and when you're comfortable, feel free to move on. The next few multiplication methods (for 5, 9, and 8) are a bit more challenging, but fundamentally rely on ideas you've already learned (adding neighbors, doubling, halving, and the like).

Quick Review

1. **The Trachtenberg System**: The **Trachtenberg System** provides the means to solve certain small-number multiplication problems quickly and efficiently. The **Trachtenberg System** methods for multiplying *n* by the numbers eleven, twelve, six, and seven are worth learning due to their simplicity and real-world applicability.

2. **Trachtenberg Multiplication by Eleven**: To multiply any number by eleven using the **Trachtenberg System**, imagine 0s at the beginning and end (left and right side) of the number. Starting with the rightmost digit (that is, the *actual* rightmost digit, *not* the added zero), *add each digit to its "neighbor"* (the digit to its right).

3. **Trachtenberg Multiplication by Twelve**: To multiply any number by twelve using the **Trachtenberg System**, first imagine 0s at the beginning and end of the number. Starting with the rightmost digit (the *actual* rightmost digit, not the added zero), *double each digit* and *add the neighbor*.

4. **Trachtenberg Multiplication by Six**: To multiply any number by six using the **Trachtenberg System**, first imagine 0s at the beginning and end of the number. Starting with the rightmost digit (the *actual* rightmost digit, not the added zero), *add five to a digit if the digit is odd* (keep the original digit if it's even) and *add half the neighbor*. Drop fractions from any digit that doesn't split cleanly in half.

5. **Trachtenberg Multiplication by Seven**: To multiply any number by seven using the **Trachtenberg System**, imagine 0s at the beginning and end of the number. Starting with the rightmost digit (the *actual* rightmost digit, not the added zero),

double each digit, add five if it is odd, and *add half the neighbor.*
Drop fractions from any digit that doesn't split cleanly in half.

Lesson 4: Trachtenberg Multiplication—Multiplying by 5, 9, and 8

You now know how to multiply any number—regardless of size—by 11, 12, 6, and 7 using the **Trachtenberg System**. In this lesson, you'll learn to employ similar principles to multiply numbers by 5, 9, and 8. Each concept may seem more difficult than the last, but with time and practice they'll become natural.

Multiplying by 5

To multiply *n* by five, use *five if the digit is odd* or *zero if the digit is even (or zero)*, plus *half the neighbor*. Note that this different than multiplication by six, where the five is *added* to the number if it's odd; rather, simply *replace* the digit with five if is odd, and *replace* it with zero if it is even. Let's begin with an example: $1,247 \times 5$.

01,24**70**: 5 (replace [odd] 7 with 5) + (1/2 of 0) = 5
01,2**4**70: 0 (replace [even] 4 with 0) + (1/2 of 7) = 3
01,**2**470: 0 (replace [even] 2 with 0) + (1/2 of 4) = 2
0**1**,2470: 5 (replace [odd] 1 with 5) + (1/2 of 2) = 6
01,2470: 0 + (1/2 of 1) = 0

Answer: $1,247 \times 5 = 6,235$.

This concept may be difficult to understand at first, but with practice these problems move more quickly.

You may notice a shortcut here: The rightmost digit of any number being multiplied by 5 will either be even or odd. If this digit is odd, the first (rightmost) answer digit will be a 5; otherwise, it will be a zero. It will never turn out to be anything else. Mathematically, this makes sense because any multiple of 5 will either end in 0 or 5. Adding half of the zero to the right of the rightmost (actual) digit will not change this. This shortcut can help speed up the first step of $n \times 5$ problems.

Let's do another problem together: $23,692 \times 5$.

023,69**20**: 0 + (1/2 of 0) = 0
023,6**9**20: 5 + (1/2 of 2) = 6
023,**6**920: 0 + (1/2 of 9) = 4
02**3**,6920: 5 + (1/2 of 6) = 8
0**2**3,6920: 0 + (1/2 of 3) = 1
023,6920: 0 + (1/2 of2) = 1

Answer: $23,692 \times 5 = 118,460$.

You may have noticed there's no need to carry when multiplying by 5 with this method. The largest individual computation you could come across in such a problem would be [even number] + 9, which would translate to 5 + 4, which equals 9. As a result, $n \times 5$ problems can be solved quickly once you're used to this method.

Perform the following calculations; the answers are below. Don't cheat yourself; figure out and write down the answers, then check them. Only write down the answer digits, never the problem steps or work. If you've made a mistake, try to figure out where you went wrong.

1. $3,354 \times 5$
2. $741,483 \times 5$
3. $9,103,203 \times 5$
4. $39,921,461 \times 5$

Answers:

1. 16,770
2. 3,707,415
3. 45,516,015
4. 199,607,305

Multiplying by 9

Multiplication by nine is where things become more "processor intensive." However, like anything, with practice and patience, it can become second nature. Don't be discouraged if the answers come slowly at first; until you're a mathematical wizard, they certainly will.

To multiply a number by nine, first *subtract the rightmost digit from ten* (step 1). Starting at the next digit (second from the right) and moving from right to left, *subtract each remaining digit from nine* and *add its neighbor* (step 2). Finally, once you've applied step 2 to the leftmost digit and its neighbor, *subtract one from the leftmost digit* (step 3).

"Wait...What?"

Let's walk through an example to illustrate the process: 283 × 9.

Normally,

283 × 9

...becomes...

0283<u>0</u> × 9

...but in this case, you can ignore this step, as the calculations applied to the right- and left-most digits don't require use of their neighbors. First, *subtract the rightmost digit from ten* (step 1):

Step 1: 28<u>3</u>: $(10 - 3) = 7$

Starting at the next digit (second from the right) and moving from right to left, *subtract each remaining digit from nine* and *add its neighbor*:

Step 2: 2<u>8</u>3: $(9 - 8) + 3 = 4$
Step 2 (continued): <u>2</u>83: $(9 - 2) + 8 = 5$ (carry the 1)

Finally, once you've applied step 2 to the leftmost digit and its neighbor, *subtract one from the leftmost digit*:

Step 3: <u>2</u>83: (2 − 1) + (the carried) 1 = 2

Answer: 283 × 9 = 2,547.

If that didn't make sense, go through the problem again slowly, and refer to the instructions during each step. Once you think you have it, try another example: 3,936 × 9. Again, break it down step-by-step.

Step 1:

3,93<u>6</u>: (10 − 6) = 4

Step 2:

3,9<u>3</u>6: (9 − 3) + 6 = 2 (carry the 1)
3,<u>9</u>36: (9 − 9) + 3 + (the carried) 1 = 4
<u>3</u>,936: (9 − 3) + 9 = 5 (carry the 1)

Step 3:

<u>3</u>,936: (3 − 1) + (the carried) 1 = 3

Answer: 3,936 × 9 = 35,424.

You may ask, "Is this truly easier than classic math?" Have you tried doing a problem like 41,149 × 9 in your head recently? Using classical methods, it wouldn't be easy. When it's time to move calculation off-paper and into your brain, this will be your go-to system for precise small number multiplication. It only seems daunting now because it involves a few unfamiliar steps. Comfort will come with time and practice.

Perform the following calculations; the answers are below. Don't cheat yourself; figure out and write down the answers, then check them. Only write down the answer digits, never the problem steps or work. If you've made a mistake, try to figure out where you went wrong.

1. $3{,}144 \times 9$
2. $910{,}123 \times 9$
3. $223{,}451 \times 9$
4. $99{,}927{,}512 \times 9$

Answers:

1. 28,296
2. 8,191,107
3. 2,011,059
4. 899,347,608

Multiplying by 8

On its own, multiplication by eight is relatively complex. Having learned to multiply by nine first will ease the process somewhat as the two methods share many core concepts. That said, eight is by far the most difficult **Trachtenberg System** individual multiplication method you'll learn.

If your head is swimming, spend some more practice time with nine before tackling eight.

To multiply a number by eight, first *subtract the rightmost digit from ten* and *double the outcome* (step 1). Next, moving from right to left (like with nine), *subtract each remaining digit from nine.* Unlike nine, *double the outcome* (step 2) before adding the neighbor. Finally, *subtract two from the leftmost digit* (step 3).

Relax. It sounds harder than it is. Time for an example: 346 × 8.

Step 1 (like 9, you don't need to "bookend" it with zeroes):

34**6**: [2 × (10 − 6)] = 8

Step 2:

3**4**6: [2 × (9 − 4)] + 6 = 6 (carry the 1)
346: [2 × (9 − 3)] + 4 + (the carried) 1 = 7 (carry the 1)

Step 3:

346: (3 − 2) + (the carried) 1 = 2

Answer: 346 × 8 = 2,768.

Were you able to keep up? Go through it one more time, referring to the instructions as needed.

Let's do one more together: 1,291 × 8.

Step 1:

1,29**1**: [2 × (10 − 1)] = 8 (carry the 1)

Step 2:

1,2**9**1: [2 × (9 − 9)] + 1 + (the carried) 1 = 2

1,**29**1: $[2 \times (9 - 2)] + 9 = 3$ (carry the 2)
1,291: $[2 \times (9 - 1)] + 2 +$ (the carried) $2 = 0$ (carry the 2)

Step 3:

1,291: $(1 - 2) +$ (the carried) $2 = 1$

Answer: $1{,}291 \times 8 = 10{,}328$.

You may have noticed something about the last step of the $n \times 8$ method; step 3 ("subtract 2 from the leftmost digit") seems to imply the possibility of a negative answer digit.

The short story: don't worry about it; it will never happen.

The long story: let's walk through an example to demonstrate why it won't be a problem. We'll use 199×8.

Step 1:

19**9**: $[2 \times (10 - 9)] = 2$

Step 2:

1**99**: $[2 \times (9 - 9)] + 9 = 9$
199: $[2 \times (9 - 1)] + 9 = 5$ (carry the 2)

Step 3:

199: $(1 - 2) +$ (the carried) $2 = 1$

Answer: $199 \times 8 = 1{,}592$.

As you can see, the presence of a 1 in the leftmost place of n (1̲99)—the one which seems to pose the risk of a negative digit—is the very same thing that prevents a negative digit from happening. In the previous step, this 1 was acted upon like so: $[2 \times (9 - 1)]$. This at minimum (not even considering the neighbor) produces 16, forcing you to carry a 1 into the "risky" step.

If that was confusing, stick with the short story (don't worry about it; it will never happen).

As laborious as $n \times 8$ problems may seem at first, there's no catch to them; comfort and speed will come with practice. Perform the following calculations; the answers are below. Don't cheat yourself; figure out and write down the answers, then check them. Only write down the answer digits, never the problem steps or work. If you've made a mistake, try to figure out where you went wrong.

1. $2,164 \times 8$
2. $813,611 \times 8$
3. $422,469 \times 8$
4. $43,927,610 \times 8$

Answers:

1. 17,312
2. 6,508,888
3. 3,379,752
4. 351,420,880

Goodbye to Trachtenberg...For Now.

This is the end of our study of **Trachtenberg System** methods for multiplication by specific, small, practical numbers. Using this system, numbers *can* be multiplied by 2, 3 and 4, but the methods are absurdly complex (more so than the methods for multiplication by 8 and 9). This book is about *practical* math, and so I couldn't in good conscious include them.

Should you wish to pursue **Trachtenberg System** methods for 2, 3, and 4, your morbid curiosity shall not go unfulfilled. There are many resources available on the Internet, as well as the aforementioned Cutler/McShane publication, *The Trachtenberg System of Speed Arithmetic*.

With time, you'll learn to use the right tool for each multiplication challenge you face. You wouldn't go to war with a can of bug spray, and you wouldn't attack ants in the kitchen with a 50-caliber machine gun. Similarly, you don't need the powerful **Trachtenberg System** to solve 12 × 19, because it's faster and easier to solve it with the **Teeny-Teeny Shortcut**. Before blindly attempting to use any of the several multiplication methods you've learned, it's critical to be able to recognize immediately which method is best for each case.

Take a few days to work on all the **Trachtenberg System** multiplication strategies you've learned; all in all, this means the methods for multiplying by 11, 12, 6, 7, 5, 9, and 8. Focus on the ones you learned most recently, and practice as often as you can. There's no race to finish *Master the Language of the Universe*; your challenge is simply to better yourself at a reasonable pace. There will be further reinforcement and review at the conclusion of this section, and with time, you'll learn to handle these problems in your mind, using no external tools.

Quick Review

1. **The Trachtenberg System**: The **Trachtenberg System** provides the means to solve certain small-number multiplication problems quickly and efficiently. The **Trachtenberg System** methods for multiplying *n* by the numbers five, nine and eight are worth learning due to their relative simplicity and real-world applicability.

2. **Trachtenberg Multiplication by Five**: To multiply any number by five using the **Trachtenberg System**, first insert imaginary 0s at the beginning and end (left and right side) of the number. Starting with the rightmost digit (the *actual* rightmost digit, not the added zero), use *five if the digit is odd* or *zero if the digit is even (or zero)*, plus *half the neighbor*. Keep in mind that this is different than the *n* × 6 method, where the five is *added to the number* if it's odd. Here, it isn't added, but instead *the five simply replaces the digit* if it's odd. If it's even, *replace the digit with zero*.

3. **Trachtenberg Multiplication by Nine**: To multiply a number by nine using the **Trachtenberg System**, first *subtract the rightmost digit from ten* (step 1). Starting at the next digit (second from the right) and working from right to left, *subtract each remaining digit from nine* and *add the neighbor* (step 2). Finally, once you've applied step 2 to the last (leftmost) digit and its neighbor, *subtract one from the leftmost digit* (step 3).

4. **Trachtenberg Multiplication by Eight**: To multiply any number by eight using the **Trachtenberg System**, first *subtract*

the rightmost digit from ten and *double the outcome* (step 1). Next, moving right to left, *subtract each remaining digit from nine*. Unlike the method for $n \times 9$, *double the outcome before adding the neighbor* (step 2). Finally, *subtract two from the leftmost digit* (step 3).

Review and Development: Section 3

In *Section 3*, you learned how to handle common types of practical multiplication problems using the **Trachtenberg System**, as well as a series of specific multiplication **Shortcuts**.

Review

1. <u>**The Trachtenberg System**</u>: The **Trachtenberg System** provides the means to solve certain small-number multiplication problems quickly and efficiently, requiring case-specific and otherwise-unused skills. The **Trachtenberg System** methods for multiplying *n* by the numbers five, six, seven, eight, nine, eleven, and twelve are worth learning due to their relative simplicity and real-world applicability.

2. <u>**Trachtenberg Multiplication by Eleven**</u>: To multiply any number by eleven using the **Trachtenberg System**, imagine 0s at the beginning and end (left and right side) of the number. Starting with the rightmost digit (that is, the *actual* rightmost digit, *not* the added zero), *add each digit to its "neighbor"* (the digit to its right).

3. <u>**Trachtenberg Multiplication by Twelve**</u>: To multiply any number by twelve using the **Trachtenberg System**, first imagine 0s at the beginning and end of the number. Starting with the rightmost digit (the *actual* rightmost digit, not the added zero), *double each digit* and *add the neighbor*.

4. <u>**Trachtenberg Multiplication by Six**</u>: To multiply any number by six using the **Trachtenberg System**, first imagine 0s at the beginning and end of the number. Starting with the rightmost

digit (the *actual* rightmost digit, not the added zero), *add five to a digit if the digit is odd* (keep the original digit if it's even) and *add half the neighbor*. Drop fractions from any digit that doesn't split cleanly in half.

5. **Trachtenberg Multiplication by Seven**: To multiply any number by seven using the **Trachtenberg System**, imagine 0s at the beginning and end of the number. Starting with the rightmost digit (the *actual* rightmost digit, not the added zero), *double each digit, add five if it is odd*, and *add half the neighbor*. Drop fractions from any digit that doesn't split cleanly in half.

6. **Trachtenberg Multiplication by Five**: To multiply any number by five using the **Trachtenberg System**, first insert imaginary 0s at the beginning and end (left and right side) of the number. Starting with the rightmost digit (the *actual* rightmost digit, not the added zero), use *five if the digit is odd* or *zero if the digit is even (or zero)*, plus *half the neighbor*. Keep in mind that this is different than the $n \times 6$ method, where the five is *added to the number* if it's odd. Here, it isn't added, but instead *the five simply replaces the digit* if it's odd. If it's even, *replace the digit with zero*.

7. **Trachtenberg Multiplication by Nine**: To multiply a number by nine using the **Trachtenberg System**, first *subtract the far right digit from ten* (step 1). Starting at the next digit (second from the right) and working from right to left, *subtract each remaining digit from nine* and *add the neighbor* (step 2). Finally, once you've applied step 2 to the last (leftmost) digit and its neighbor, *subtract one from the leftmost digit* (step 3).

8. **Trachtenberg Multiplication by Eight**: To multiply any number by eight using the **Trachtenberg System**, first *subtract the rightmost digit from ten* and *double the outcome* (step 1). Next, moving right to left, *subtract each remaining digit from*

nine. Unlike the method for $n \times 9$, *double the outcome before adding the neighbor* (step 2). Finally, *subtract two from the leftmost digit* (step 3).

9. **Multiplication Shortcuts**: There are several situation-specific Shortcuts for solving practical multiplication problems. These include **Teeny-Teeny, Splitsy-Doubly, Double Double, the Five-Ten Split, Nine to Ten and Down Again,** the **Quarter Pounder, the Jigsaw, the Double Down,** and **FOIL**.

The **Trachtenberg System** methods taught in this section will surely take a while to get down, as you have two major obstacles to overcome. First, you must instantly recall the specific **Trachtenberg System** strategy for each individual number ("Multiply by 12? Ummm, double the number, and add the neighbor"), and second, you must become quick enough at implementing each strategy that you're able to apply them reasonably.

In order to practice, you need to get into the habit of multiplying numbers found all around you by the numbers five through twelve. When doing so, if a number you find is less than 999, try to remember the problem and answer digits in your head; otherwise, feel free to write them down (for now).

When I was first learning the **Trachtenberg System** methods for the specific numbers discussed, I vividly recall having to sit through a very long Catholic wedding ceremony. Due to some missteps on the part of the organizers, it turned out to be about double the length of a normal wedding. As someone with a relatively short attention span, I felt as though I was being tortured. On a dark wooden panel at the front of the church, I could see a list of all the hymns that were to be sung throughout the service; essentially, it was a list containing about eight three-digit numbers. I spent the entire mass staring at this panel, honing my **Trachtenberg System** multiplication skills and discreetly scribbling down and checking my answers on the back of the wedding program.

The point of this anecdote is that numbers are everywhere, ready to be multiplied. The excitement and pride that accompanies your own advancement should be enough incentive to keep you practicing. Soon, you'll be faster with this type of multiplication than you may have ever thought possible.

Multiplication Shortcuts share the same need for practice; you must exercise the skills involved whenever possible in order to feel comfortable using them.

Most importantly, don't give up. Think about how long it took to learn classical math as a student, and how many years were dedicated to simple arithmetic. You (most likely) have a much more highly-developed brain than you did at that age, but these are still entirely new and foreign skills, and you should account for the time it may take to develop them accordingly.

Here's a **Trachtenberg System** cheat sheet for multiplying numbers by five through twelve:

5	Add five if the number is odd, otherwise add zero; Add half the neighbor.
6	Add five if the number is odd, otherwise add zero; Add half the neighbor.
7	Double the number, and, if it is odd, add five, otherwise add

	zero. Add half the neighbor.	
8	Subtract the rightmost digit from ten and double the outcome; Subtract the "middle" digits from nine, double the outcomes, and add the neighbors; Subtract two from the leftmost digit.	
9	Subtract the rightmost digit from ten; Subtract the "middle" digits from nine and add the neighbors; Subtract one from the leftmost digit.	
10	Add a zero to the right side of the number.	
11	Add the neighbor.	
12	Double the number; Add the neighbor.	

Development

Spend one week working on the skills learned in *Section 3*.

Day 1

Re-read *Section 3* and spend extra time on anything that seems unclear or difficult. Memorize the steps involved in each method until you develop a high degree of confidence (for both the **Trachtenberg System** and **Shortcuts**).

If someone were to ask you how to multiply a number *n* by 8 using the **Trachtenberg System**, you should be able to quickly reply, "Subtract the rightmost digit from ten and double the outcome;

subtract the 'middle' digits from nine, double the outcomes, and add the neighbors; subtract two from the leftmost digit."

If asked how to perform the **Nine to Ten and Down Again Shortcut**, you should be able to reply, "Multiply n by 10, and then subtract n from the outcome." By the end of *Day 1*, your responses should be immediate and automatic.

Days 2 – 4

Review the **Trachtenberg System** multiplication methods at least three times each day. Ideally, do so early in the morning, once mid-day, and at night before bed. Multiply numbers you encounter throughout the day by 5 through 12 using these methods.

In addition to using the world around you, multiply the following numbers by 5, 6, 7, 8, 9, 11, and 12 using the **Trachtenberg System**. The answers are listed below.

1. 1,234
2. 1,111
3. 1,894
4. 2,199
5. 2,286
6. 4,691
7. 6,770
8. 9,137
9. 9,987
10. 11,187
11. 19,674
12. 54,860

13. 89,910

14. 180,722

15. 689,632

Answers:

1. 5: 6,170; 6: 7,404; 7: 8,638; 8: 9,872; 9: 11,106; 11: 13,574; 12: 14,808

2. 5: 5,555; 6: 6,666; 7: 7,777; 8: 8,888; 9: 9,999; 11: 12,221; 12: 13,332

3. 5: 9,470; 6: 11,364; 7: 13,258; 8: 15,152; 9: 17,046; 11: 20,834; 12: 22,728

4. 5: 10,995; 6: 13,194; 7: 15,393; 8: 17,592; 9: 19,791; 11: 24,189; 12: 26,388

5. 5: 11,430; 6: 13,716; 7: 16,002; 8: 18,288; 9: 20,574; 11: 25,146; 12: 27,432

6. 5: 23,455; 6: 28,146; 7: 32,837; 8: 37,528; 9: 42,219; 11: 51,601; 12: 56,292

7. 5: 33,850; 6: 40,620; 7: 47,390; 8: 54,160; 9: 60,930; 11: 74,470; 12: 81,240

8. 5: 45,685; 6: 54,822; 7: 63,959; 8: 73,096; 9: 82,233; 11: 100,507; 12: 109,644

9. 5: 49,935; 6: 59,922; 7: 69,909; 8: 79,896; 9: 89,883; 11: 109,857; 12: 119,844

10. 5: 55,935; 6: 67,122; 7: 78,309; 8: 89,496; 9: 100,683; 11: 123,057; 12: 134,244

11. 5: 98,370; 6: 118,044; 7: 137,718; 8: 157,392; 9: 177,066; 11: 216,414; 12: 236,088

12. 5: 274,300; 6: 329,160; 7: 384,020; 8: 438,880; 9: 493,740; 11: 603,460; 12: 658,320

13. 5: 449,550; 6: 539,460; 7: 629,370; 8: 719,280; 9: 809,190; 11: 989,010; 12: 1,078,920

14. 5: 903,610; 6: 1,084,332; 7: 1,265,054; 8: 1,445,776; 9: 1,626,498; 11: 1,987,942; 12: 2,168,664

15. 5: 3,448,160; 6: 4,137,792; 7: 4,827,424; 8: 5,517,056; 9: 6,206,688; 11: 7,585,952; 12: 8,275,584

If you would like even more practice opportunities, feel free to take more than two days.

Days 5 – 7

On days 5 through 7, focus on **Multiplication Shortcuts**. Be sure you know when and how to apply each **Shortcut**.

Practice with the problems listed below. Approach them in two phases. Don't try to figure out the answers at first; instead, practice quickly identifying the best method for solving each problem. Next, solve the problem using that method. The answers are below. Keep in mind there are sometimes multiple methods that could apply. Based on personal preference, you may prefer to use **FOIL** on a problem that could be solved with the **Teeny-Teeny** or **Jigsaw** methods. For this reason, **FOIL** has been omitted from the answers, but remember that it's always a great option for multiplying two two-digit numbers.

1. 12×44
2. 34×11
3. 4×26
4. 71×8
5. 18×5
6. 168×20

7. 9×44

8. 12×366

9. 19×4

10. $1{,}333 \times 5$

11. 55×55

12. 18×14

13. 41×49

14. 21×8

15. 48×20

16. 160×25

17. 239×5

18. 221×11

19. 33×37

20. 289×5

21. 13×18

22. 19×64

23. 25×580

24. 318×8

25. 9×648

26. 9×18

27. 7×31

28. 208×5

29. 6×121

30. 3×32

Answers (and best methods):

1. 528 – Trachtenberg 12

2. 374 – Trachtenberg 11

3. 104 – Double-Double

4. 568 – Trachtenberg 8

5. 90 – The Five-Ten Split
6. 3,360 – The Double-Down
7. 396 – Nine to Ten and Down Again (or Trachtenberg 9, though it's a manageable number).
8. 4,392 – Trachtenberg 12
9. 76 – Double-Double
10. 6,665 – Trachtenberg 5
11. 3,025 – The Jigsaw
12. 252 – Teeny-Teeny
13. 2,009 – The Jigsaw
14. 168 – Trachtenberg 8
15. 960 – The Double-Down
16. 4,000 – The Quarter Pounder
17. 1,195 – Trachtenberg 5
18. 2,331 – Trachtenberg 11
19. 1,221 – The Jigsaw
20. 1,445 – The Five-Ten Split
21. 234 – Teeny-Teeny
22. 1,216 – Splitsy-Doubly
23. 14,500 – The Quarter Pounder
24. 2,544 – Trachtenberg 8
25. 5,832 – Trachtenberg 9
26. 162 – Nine to Ten and Down Again
27. 217 – Trachtenberg 7
28. 1,040 – The Five-Ten Split
29. 726 – Trachtenberg 6
30. 96 – Splitsy-Doubly

While focusing on the most recent lessons, continue working to further develop all the skills you've learned so far in *Master the*

Language of the Universe. Next, you'll learn some **Shortcuts** and **Trachtenberg System** methods for division.

Section 4: New Approaches to Division

It's been a wild ride through the bumpy back roads of mathematics, but we're just getting started. It's time to address division.

You gasped, didn't you?

You use division all the time, but it can be more difficult than adding, subtracting, or multiplying because it doesn't translate as readily to life experience. Imagine these (delicious) scenarios:

- You have before you a slice of cake (for reference), and you must shut your eyes and imagine six and a half slices.
- You have before you a whole cake (for reference), and you must imagine it split into six and a half slices.

The first is less mentally taxing. Because you have the individual slice of cake as a frame of reference, you only have to imagine six of them, along with a final slice half the size of the others.

The second scenario is a bit more complex, because you have to imagine the entire reference unit (one whole cake) split into six and a half parts. Most people would instinctively try to imagine the cake split into six slices, and then imagine the removal of a sliver from each slice to contribute to the half slice. Technically, constructing a half slice by taking an equal part from each of the six slices would require you to remove approximately 9.2% from each slice, so it would be quite difficult to imagine this with any precision.

Division intimidates people because it's often difficult to gracefully relate objects to even simple division problems (as you just saw); however, it's important to realize that being able to divide quickly and accurately is a skill that can greatly improve your overall ability to solve problems, mathematical or otherwise.

Let's get right into it.

Lesson 1: Two Quick Division Shortcuts

There is certainly a relationship between freedom and mathematics. Learning multiple ways to solve different problem types expands your options; this knowledge gives you the freedom to take different roads to the same destination.

Unlike multiplication, there aren't many practical division **Shortcuts** worth learning at this time. However, to ease you into division concepts, you'll learn to handle two common division problem types using **Shortcuts**. These two types are "any number divided by five" (*n*/5) and "any number divided by four" (*n*/4).

The Two-and-Shift

To divide any number by 5, perform this simple two-step operation:

1. Multiply the number by 2, then

2. Move the decimal place one space to the left.

This can be expressed algebraically as $n/5 = 0.1(2n)$. Let's look at an example: $33 \div 5$.

---
1. $33 \times 2 = 66$
2. $66 \times 0.1 = 6.6$

Answer: 6.6.

Do you see how that worked? Let's do another example: $56 \div 5$.

---
1. $56 \times 2 = 112$
2. $112 \times 0.1 = 11.2$

Answer: 11.2.

The next example is a bit more difficult, due to the size of the n: $1,882 \div 5$.

---
1. $1,882 \times 2 = 3,764$
2. $3,764 \times 0.1 = 376.4$

Answer: 376.4.

Above, you took the time to precisely multiply $1,882 \times 2$, but if the context of the problem requires only a rough answer, you could approach it this way instead:

---
1. $1,882 \times 2 = (1,900 \times 2) = 3,800$
2. $3,800 \times 0.1 = 380$

This results in an answer of "around 380." The actual answer is 376.4, so the rough calculation is clearly in the ballpark and would be appropriate for many non-scientific applications.

Let's do one more. In this example, n involves a decimal point: $36.8 \div 5$. Do a rough calculation first, then again with precision.

Rough:

1. $36.8 \times 2 = (37 \times 2) = 74$
2. $74 \times 0.1 = 7.4$

Precise:

1. $36.8 \times 2 = 73.6$
2. $73.6 \times 0.1 = 7.36$

Perform the following calculations; the precise answers are below. Don't cheat yourself; figure out and write down the answers, then check them. Only write down the answer digits, never the problem steps or work. If you've made a mistake, try to figure out where you went wrong.

1. $61 \div 5$
2. $39 \div 5$
3. $57 \div 5$
4. $66.6 \div 5$
5. $121.6 \div 5$
6. $222 \div 5$
7. $602 \div 5$
8. $2,088 \div 5$

Answers:

1. 12.2
2. 7.8
3. 11.4
4. 13.32
5. 24.32
6. 44.4
7. 120.4
8. 417.6

Half-and-Half

To divide any number by 4, perform this simple two-step operation:

1. Cut the number in half, and
2. Cut it in half again.

It seems obvious, but it's important to remember this is an option for $n/4$ problems. Often, the perceived pressure you feel to solve problems quickly can give you tunnel vision.

Expressed algebraically, this is $n/4 = (n/2)/2$.

Let's begin with an easy example: $460 \div 4$.

1. $460 \div 2 = 230$
2. $230 \div 2 = 115$

Answer: 115.

Now let's try one that doesn't **Split** so gracefully: $316 \div 4$.

1. $316 \div 2 = 158$
2. $158 \div 2 = 79$

Answer: 79.

Going a step further, try an example that involves an odd number: $121 \div 4$.

1. $121 \div 2 = 60.5$
2. $60.5 \div 2 = 30.25$

Answer: 30.25. As with the **Two-and-Shift**, there is the option of **Splitting** more loosely to arrive at slightly less precise answers:

1. $121 \div 2 = 120 \div 2 = 60$
2. $60 \div 2 = 30$

The answer is "about 30." If only marginally, $120 \div 2$ is easier to solve than $121 \div 2$, and the difference between the precise answer and rough answer is 0.25. In most practical cases, you'll be dealing with small numbers like these, and the precision you're sacrificing by rounding is often insignificant.

Much like the **Splitsy-Doubly** multiplication method you learned in the last section, this works especially well with powers of two, and is considerably more difficult if n is odd.

Perform the following calculations; the precise answers are below. Don't cheat yourself; figure out and write down the answers, then check them. Only write down the answer digits, never the problem steps or work. If you've made a mistake, try to figure out where you went wrong.

1. $14 \div 4$
2. $39 \div 4$
3. $70 \div 4$
4. $118 \div 4$
5. $3,180 \div 4$

Answers:

1. 3.5
2. 9.75
3. 17.5
4. 29.5
5. 795

We'll stop here. Though we only covered two division **Shortcuts**; hopefully this got you into a dividing mindset. You'll be able to come up with **Shortcuts** of your own using the next technique, the *Fraction-to-Decimal (F2D) Method*. This will expand your division vocabulary and drastically increase your comfort level with many common division problems.

The best tools you can have in your division toolbox are *comfort* and *confidence*, which will increase the more you work with division in general.

Quick Review

1. Becoming familiar with and using a pair of simple division **Shortcuts** (the **Half-and-Half** and the **Two-and-Shift**) allows you to more easily address some common division problems.

Lesson 2: The Fraction-to-Decimal (F2D) Method

A perfect number is by definition a positive integer that is equal to the sum of its positive divisors. Based on this definition, 1 is not a perfect number, but in a less scientific sense, it could be argued that 1 is a *conceptually* perfect number. It has an intrinsic beauty and simplicity. It's easily imagined, understood, communicated, and manipulated. Developing children can grasp the concept of 1.

In the same way, it could also be argued that 0.142857 is conceptually imperfect; it is difficult to imagine, difficult to memorize, unfamiliar, and contains no repeated digits. You'll deal with this number in this lesson: 0.142857 equals 1/7. Here, you'll explore common fractions, their decimal counterparts, and how to leverage their relationships to better understand division.

Knowing how to convert fractions into decimals quickly will allow you to:

- Immediately (without calculations) solve division problems involving small numbers divided by other small numbers;
- Immediately translate "ugly" numbers (like 0.142857) into language that is easy to communicate and understand; and
- Quickly convert the remainders of many larger division problems into precise decimal values.

On the first point, imagine you just pulled off an *Oceans Eleven*-style heist and now have to divide one million dollars among six people. Using the **F2D Method**, you'd know that 1 ÷ 6 is .166666, so each individual is entitled to about $166,666. Or, say you need to ask eight people which television show they most enjoyed from the 1991-92 season of ABC's Friday Night *TGIF* lineup. Of the eight individuals polled, five of them (5/8) choose the same show (*Family Matters*). What percentage is that? You could do this using math learned earlier in this book, but after this lesson, there will be no need for calculation; you'll simply know that the answer is 62.5%.

The second point refers to the human tendency to speak and conceptualize numbers both as decimals and percentages or pieces of a whole. It's common knowledge that 1/2 of something is the same as 50% or 0.5. Can you as easily say the same for 1/8 or 2/7? If you found out you lost 28.6% of your wealth in a stock market crash, would you know that that this translates to about 2/7?

On the third point: what's 51 divided by 7? You probably couldn't quickly produce the exact answer (7.285714), but using your knowledge of classical division and multiplication tables, you'd figure out fairly quickly that 7 goes into 51 about 7 times (as 7 × 7 equals 49). You'd also determine that there's a remainder of 2 (as 51 − 49 equals 2). Some situations might call for the precise decimal answer as opposed to the mixed number, "7 and 2/7." Could you convert "7 and 2/7" into 7.285714 without a calculator or other external aid? Most of us can't. Here you'll learn how to quickly come up with precise decimal answers for many common division problems that leave you with single-digit remainders.

> ***Nerdy Fact***: *The "division" sign (÷) is actually called an "obelus."*

The Fraction Groups

You know by now that every fraction has a decimal equivalent; "one half" is the same as 0.5, "one tenth" is the same as 0.1, etc. Even fractions like "one third" have decimal equivalents, even though they may not necessarily terminate (0.333333...). When dividing any reasonably small number by a single-digit number, you can turn an answer with a remainder into a precise decimal answer by memorizing a few **Fraction-to-Decimal (F2D)** conversions and learning to apply the patterns they demonstrate. Even just memorizing the decimal equivalent of every fractional combination up to 10 will give you a drastic advantage with most day-to-day division. That's precisely what you'll do.

To begin, let's look at your first basic set of fractions, which you'll refer to as the *1/n* **Fraction Group**.

1/n Fraction Group

- 1/2 = 0.5
- 1/3 = 0.333...

- 1/4 = 0.25
- 1/5 = 0.2
- 1/6 = 0.166
- 1/7 = 0.142857
- 1/8 = 0.125
- 1/9 = 0.111...
- 1/10 = 0.1

For most people, the conversions listed above fall into two groups. Some of the precise decimal equivalents are likely familiar, like 1/2, 1/3, 1/4, 1/5 and 1/10; others are likely unfamiliar, like 1/6, 1/7, 1/8, and 1/9. Once you've memorized these conversions, you can apply this knowledge to related fractions. For example, 2/3 = (1/3 × 2) = (0.333333... × 2) = 0.666666. As another example, 3/8 = (1/8 × 3) = (0.125 × 3) = 0.375.

Let's continue with the *2/n* fraction group and move on immediately from there. Don't just memorize the below tables, but rather try to recognize patterns and make sense of the logic behind each relationship; that's where the power lies.

You'll see a note next to some of the more daunting conversions you're about to learn: "see 'Some **F2D** Shortcuts.'" In these cases, don't attempt to memorize through brute force; read on and you'll soon learn tricks that will make them significantly easier to commit to memory.

2/n Fraction Group

- 2/3 = 0.666...
- 2/4 = 0.5
- 2/5 = 0.4 (see "Some **F2D** Shortcuts")

- 2/6 = 0.333...
- 2/7 = 0.285714 (see "Some **F2D** Shortcuts")
- 2/8 = 0.25 (see "Some **F2D** Shortcuts")
- 2/9 = 0.222...
- 2/10 = 0.2

3/n Fraction Group

- 3/4 = 0.75
- 3/5 = 0.6 (see "Some **F2D** Shortcuts")
- 3/6 = 0.5
- 3/7 = 0.428571 (see "Some **F2D** Shortcuts")
- 3/8 = 0.375 (see "Some **F2D** Shortcuts")
- 3/9 = 0.333...
- 3/10 = 0.3

4/n Fraction Group

- 4/5 = 0.8 (see "Some **F2D** Shortcuts")
- 4/6 = 0.666...
- 4/7 = 0.571428 (see "Some **F2D** Shortcuts")
- 4/8 = 0.5 (see "Some **F2D** Shortcuts")
- 4/9 = 0.444...
- 4/10 = 0.4

5/n Fraction Group

- 5/6 = 0.8333...
- 5/7 = 0.714285 (see "Some **F2D** Shortcuts")
- 5/8 = 0.625 (see "Some **F2D** Shortcuts")

- $5/9 = 0.555...$
- $5/10 = 0.5$

6/n Fraction Group

- $6/7 = 0.857142$ (see "Some F2D Shortcuts")
- $6/8 = 0.75$ (see "Some F2D Shortcuts")
- $6/9 = 0.666...$
- $6/10 = 0.6$

7/n Fraction Group

- $7/8 = 0.875$
- $7/9 = 0.777...$
- $7/10 = 0.7$

8/n Fraction Group

- $8/9 = 0.888...$
- $8/10 = 0.8$

9/n Fraction Group

- $9/10 = 0.9$

Some F2D Shortcuts

Before you fully memorize these tables, let's look at some
Shortcuts.

The Familiar Seven Pattern

While learning the **Fraction Groups** (*1/n*, *2/n*, *3/n*, etc.), you
may have noticed something about *n/7* (*1/7, 2/7, 3/7*, etc.); they're a
pain in the ass. Here's a trick you can use to memorize them.

First, memorize the integer 142857 (you may recognize this from
the conversion tables; 1/7 = 0.142857). Throughout the *n/7*'s, the
number 142857 is repeated, but it "begins" at different points.

- 1/7 = 0.142857 (begins on the 1 of 142857)
- 2/7 = 0.285714 (begins on the 2 of 142857)
- 3/7 = 0.428571 (begins on the 4 of 142857)
- 4/7 = 0.571428 (begins on the 5 of 142857)
- 5/7 = 0.714285 (begins on the 7 of 142857)
- 6/7 = 0.857142 (begins on the 8 of 142857)

Memorizing this pattern removes the perceived randomness
from the *n/7* fractions, making them easier to tackle. Those of you with
classical musical training can liken this to modes; to figure out different
modes (D Dorian, E Phrygian, F Lydian, etc.), one would use the notes
of the C major scale (Ionian), but begin and end the octave at different
points within it.

Eight's "Quartery" Style

Technically, you can convert the *n*/8 fractions by multiplying the numerator (or "top" number) by 125 and moving the decimal point to the left of the product. However, that isn't easy to perform on the fly, and the point of this whole method is "easy," right? Instead, you can simply memorize the following list. As you'll see, all the *n*/8 factions have decimal equivalents that end in "quarters" (25, 5, 75).

- 1/8 = 0.125
- 2/8 = 0.25
- 3/8 = 0.375
- 4/8 = 0.5
- 5/8 = 0.625
- 6/8 = 0.75
- 7/8 = 0.875

The Repetitive Nines

The *n*/9 fractions couldn't be easier; just repeat the numerator.

- 1/9 = 0.111...
- 2/9 = 0.222...
- 3/9 = 0.333...
- 4/9 = 0.444...
- 5/9 = 0.555...
- 6/9 = 0.666...
- 7/9 = 0.777...
- 8/9 = 0.888...

The Fives: Multiply by Two and Move the Decimal Place

Remember the **Two-and-Shift** division **Shortcut**? The *n/5* fraction group can be translated to decimal notation by multiplying the numerator by 2 and moving the decimal place one space to the left. For example, **3**/5 = (**3** × 2 = 6; move the decimal place: 0.6).

- 1/5 = 0.2
- 2/5 = 0.4
- 3/5 = 0.6
- 4/5 = 0.8

This trick even works for fractions that translate into decimal numbers that are larger than 1 (that is to say, larger than 5/5). For example, **8**/5 = (**8** × 2 = 16; move the decimal place: 1.6). The pattern continues indefinitely:

- ...
- 12/5 = 2.4
- 13/5 = 2.6
- ...
- 64/5 = 12.8
- 65/5 = 13
- ...
- 405/5 = 81

The Elevens: Nothin' but Repeating Nines

You know your multiplication tables for 1 through 9 pretty well by now. The *n/11* **Fraction Group** is the "9" multiplication table repeated after the decimal place. Take a look:

- 1/11 = .090909...
- 2/11 = .181818...
- 3/11 = .272727...
- 4/11 = .363636...
- 5/11 = .454545...
- 6/11 = .545454...
- 7/11 = .636363...
- 8/11 = .727272...
- 9/11 = .818181...
- 10/11 = .909090...

In the case of 1/11, the single digit answer of 1 × 9 (9) is represented as two digits (place a 0 in front of 9 to make it 09).

Application

Earlier, you learned three practical applications for the **F2D Method**:

1. Immediately (without calculations) solve division problems involving small numbers divided by other small numbers,
2. Immediately translate "ugly" numbers (like 0.142857) into language that is easy to communicate and understand, and
3. Quickly convert the remainders of many larger division problems into precise decimal values.

The first two are direct applications of the **Fraction Groups** you just memorized. Let's look at the third in more detail.

Using the F2D Method to Arrive at Precise Decimal Answers

To use the **F2D Method** for basic division problems, approach the problems the same way you approached classical division in the past; first deduce exactly how many times the divisor can fit into the dividend using the *Price is Right* rule ("closest without going over").

How many times does 4 go into 61? You know 4 goes into 60 fifteen times (4 × **10** is 40 and 4 × **5** is 20). You are left with 1 (61 − 60 = 1). This remainder doesn't just mean 1, though, it actually refers to 1/4. Using the **F2D** conversions, 1/4 converts to 0.25. Since you know that 4 fits into 61 exactly 15 and 1/4 times, you can convert the fraction into a precise, calculator-like decimal answer: 15.25.

Not all division problems are that easy. More challenging situations are common: you may need to determine the number of paid vacation hours that can be used throughout a given part of the year. You may need to compute the amount of protein (in grams) needed per meal to reach a target nutritional total. You could easily find yourself facing a situation in which a three-digit number needs to be divided by a single-digit number. How would you approach 116 ÷ 9? Can the **F2D Method** even help with this?

You know that 116 is 90 + 26. You know that 9 goes into 90 exactly 10 times, and 9 goes into 26 twice, with a remainder of 8. So 10 + 2 = 12, with a remainder of 8. Thanks to the **F2D Method**, you know that 8/9 = .888... so the answer is exactly 12.888...

How about 426 ÷ 7? Using some simple, creative computation, you can determine that 7 goes into 420 exactly 60 times (7 × 6 = 42), and so you're left with a remainder of 6. What is the precise decimal answer?

From studying the **F2D Method**, you should know that 6/7 is .857142, so the answer is 60.857142.

Try one that will really make you think: 94 ÷ 16. 16 is a larger divisor than you're used to. 16 goes into 96 exactly six times, so 16 goes into 94 five times with a remainder of 14/16. You don't know how to handle 14/16 as-is, but as both 14 and 16 are even, it reduces to 7/8. Since 7/8 equals 0.875, the answer is 5.875.

Let's try a real-world example. Say you've tracked your blood pressure for 18 months and compiled the data into monthly averages. It was higher than you'd like during one of these months, but was otherwise fine. This means that you had high blood pressure during 1/18 of the time tracked. Convert this into a percentage. You know that 1/9 is 0.1111, or 11.11%. Since 1/18 is half of 1/9, you could **Split** 0.1111 to arrive at 0.0555, so your answer is 5.55%.

Now let's say your blood pressure was high 1/19 of the time. 19 doesn't share a relationship with any of the numbers discussed in this lesson, so you have to get creative. You've determined that 1/18 is 5.55% and you know that 1/20 is 5%; you can assume that 1/19 lays between these two, a little more than 5.25%. The actual answer is 5.26%.

Recognizing Patterns and Relationships

Practicing the **F2D Method** will increase your comfort with the relationships between integers, fractions, and decimals, allowing you to quickly solve fairly complex problems that involve any combination of the three. Familiarity with these relationships will begin to reveal to you the poetry inherent in mathematical patterns.

For example, where would you begin if you needed to divide a number by 125? You know that 1/8 = 0.125, so you can leverage the relationship between 8 and 125 to create your own makeshift two-step **Shortcut**. To divide any number by 125, simply multiply by 8 and move the decimal point three places to the left (accounting for the three digits in 125). Let's illustrate this with an example: 7,000 ÷ 125.

1. 7,000 × 8 = 56,000
2. 56,000 ÷ 1,000 = 56

Answer: 7,000 ÷ 125 = 56.

You could also do this in the opposite order (first divide by 1,000, then multiply by 8), depending on what's easier for the problem at hand.

Obviously, this is easier with a nice, round number like 7,000. But how about 1,823 ÷ 125? The second step of this method is easy (dividing by 1,000), but how would you solve the first step of this nasty problem (multiplying by 8)?

Jakow Trachtenberg to the rescue! You already know how to multiply any number by 8:

1. Subtract the rightmost digit from ten and double the outcome
2. Subtract the "middle" digits from nine, double the outcomes, and add the neighbors
3. Subtract two from the leftmost digit.

Following the **Trachtenberg Method**, you can determine that 1,823 × 8 = 14,584. Then move the decimal place three to the left (divide by 1,000): 14,584 ÷ 1,000 = 14.584.

This illustrates a point made at the beginning of this section; multiplication comes more naturally to us than division, so if you can turn a division problem into a series of manageable multiplication problems, do so. Combining the skills you've learned in creative ways lets you solve problems that seem daunting at first.

How would you approach $1,211 \div 625$? Get creative and string multiple tools together. You should recognize the pattern "625," as you know that $5/8 = 0.\underline{625}$. In the last example, you were working with 125, which is related to $\underline{1}/8$. You're now working with 625, which is related to $\underline{5}/8$, so you must account for that.

Think about it: You can perform the multiplication as though it was 1/8 and then divide the answer by 5. You learned to divide by five in *Division Shortcuts* (multiply the number by 2, then move the decimal one space to the left). Let's go through this problem step by step: $1,211 \div 625$.

1. $1,211 \times 8 = 9,688$
2. $9,688 \div 5 = (9,688 \times 2) \times 0.1 = 1,937.6$
3. $1,937.6 \div 1,000 = 1.9376$

Answer: $1,211 \div 625 = 1.9376$.

The problems you face won't always fall neatly in line with these patterns. You may have to divide by 634 instead of 625, for instance. If the problem calls only for a rough answer, find the nearest **F2D** pattern number and work the problem from there. To put this in perspective, consider that $1,211 \div 634 = 1.91$; by rounding the 634 to 625, the answer goes from 1.91 to 1.94, which would suffice in most practical cases.

Be creative, recognize patterns, and get comfortable with the language of mathematics. Division doesn't have to be scary. Don't be afraid to make mistakes. Play around, learn, and enjoy the challenge.

You'll learn some **Trachtenberg System** division methods next.

Quick Review

1. Learning the **Fraction-to-Decimal (F2D) Method** gives you a useful familiarity with the relationships between basic fractions and their respective decimal equivalents. You can leverage these relationships to solve specific types of division problems quickly and with calculator-like accuracy.

Lesson 3: Trachtenberg Division by 1- or 2-Digit Divisors

You want to talk about mathematics and the human spirit? Let's take a moment to appreciate the conditions under which Jakow Trachtenberg developed his elaborate and ingenious "speed math" systems.

As it turns out, he was a badass.

Born in the Ukraine, Trachtenberg breezed through school with honors. In his early twenties, he was assigned the responsibility of overseeing the navy's engineering operations. A lifelong pacifist, he fled to Germany after the Russian Revolution and worked for several years as an editor of social and scientific periodicals. The Nazis imprisoned him in a concentration camp, and he escaped. They recaptured him, but he escaped a second time. He lived out his years in Switzerland, teaching his mathematical systems at a specialized school.

You may be wondering when Trachtenberg developed his ingenious systems of **Mental Calculation**. He did it all in his head—without so much as a pen and paper—as a way to occupy his mind during his time in the concentration camps. It amazes me that something as complex as the **Trachtenberg System** could have been devised with no tools or assistance, while staring at a wall under some of the most terrifying and soul-crushing conditions a human being could possibly endure.

Trachtenberg's story is a beautiful illustration of the relationship between mathematics and the human spirit: a brilliant, innovative system was born out of complete chaos and despair. I like to think that Trachtenberg's desire to share his system with the world helped him pull through and find the courage to escape his captors.

Why Learn Trachtenberg Division?

You just spent a good amount of time learning how to handle specific types of common division problems, and yet—if you require precise answers—the vast majority of division problems remain beyond your grasp.

Read through the following facts:

1. There are exactly 3,211 wild tigers in existence;
2. They all live in Asia; and
3. Asia has a landmass of 17,139,445 square miles.

Now imagine you were asked how many *square miles of Asia per tiger* this comes to. Could you do it without a calculator or a few minutes with a pencil and paper?

Questions about tigers and landmass aren't too common, but in the real world, complex division problems can be. When they arise, you probably don't even bother trying to solve them without tools. While tools are usually fine in such cases, with a little practice, you can learn to solve them somewhat quickly in your head using the **Trachtenberg System**.

The *Trachtenberg System of Speed Mathematics* contains two methods for division: the **Simple Method** and the **Fast Method**. The **Fast Method** is much faster than the **Simple Method** (hence the name), but the **Simple Method** isn't that much simpler than the **Fast Method**. Thus, you'll learn the **Fast Method**.

Trachtenberg Division looks difficult at first. It's completely unlike the classical division methods you learned in school. Once you get the hang of it, however, you'll quickly become a pro. The method has a beautiful "shape" or "rhythm" you'll come to know very well.

You'll start by writing your solutions out by hand. After a while, all calculation will be done in your mind.

Dividing by a Two-Digit Number

Before you begin, know that you should avoid performing complex division in your head at all costs. If a problem can be simplified or doesn't require precision, take advantage. However, it's important to know how to accommodate the need for precision, should it come up.

Let's say you want to divide 7,480 by 34.

To be fair, no one ever *wants* to divide 7,480 by 34, so let's say you *need* to divide 7,480 by 34.

First, write it out:

$$7,480 \div 34$$

With any division problem that requires the **Fast Method**, first set the problem up as three distinct parts, stacked on top of one another:

1. (Top layer): The problem
2. (Middle layer): The **Working Figures**
3. (Bottom layer): The **Partial Dividends**

First take the 7 (the leftmost digit of the divisor) and drop it down into the **Partial Dividends** area. This is your first **Partial Dividend**.

7,480 ÷ 34 =	Problem
	Working Figures
7	Partial Dividends

Going forward with **Trachtenberg System** division, so you know where to look, the number you should focus on (usually the result of an action) will be underlined, while the work will not.

Note that the "action" here goes straight down from the top layer to the bottom layer. You'll see that this process follows a distinct rhythm: the action/focus goes down (as you just saw), then up and to the right, then down, then up and to the right, and so on until the problem is solved.

Next, divide this first **Partial Dividend** by the leftmost digit of the divisor (the 3 of the <u>3</u>4). With **Trachtenberg** division, always ignore any remainders (this is part of its charm). In this case, 7/3 simply equals 2. This 2 is the first (leftmost) digit of your answer.

$7,480 \div 34 = \underline{2}XX$

If you need to re-read the steps so far, please feel free. Next, this newfound answer digit (2) will be used for a few different purposes.

First, you'll multiply the 2 by the divisor (34)—but not in the way you're used to. Instead, multiply one digit at a time (first 2 × <u>3</u>, then 2 × <u>4</u>). Keep the *entire product* of the tens place calculation (2 × 3 = <u>06</u>), and only the *leftmost digit* of the product of the ones place calculation (2 × 4 = <u>0</u>8). Add these two partial outcomes together (06 + 0 = 6).

This is called the **NT Calculation**, because you keep the entire <u>n</u>umber resulting from the first part of the calculation, and only the <u>t</u>ens place digit of the second.

Subtract this **NT** number (6) from the most recent **Partial Dividend** (7). 7 − 6 = 1. This 1 is the first digit of a new number, which you'll call your first **Working Figure.** Place the working figure in the middle layer, under the next untouched digit of the dividend (the 4 in 7,4̲80):

7,480 ÷ 34 = 2	Problem
1̲	Working Figures
7 (- 6)	Partial Dividends

Going back to the "rhythm" idea, notice how the action went up and to the right? It will soon go down again as it did in the beginning.

Working Figures should always consist of two digits, and yours currently only has one (the 1 from the last step). This 1 will be the leftmost digit of the **Working Figure.** The rightmost digit is acquired by dropping the next untouched digit (the 4 in 7,4̲80) down from the dividend:

7,480 ÷ 34 = 2	Problem
14̲	Working Figures
	Partial Dividends

Let's stop here and review what you've done so far, because it can seem like a bit much:

1. Drop the 7 (first digit of the dividend) down into the **Partial Dividend** area.

2. Divide this 7 by the first digit of the divisor (3). Drop any remainders. From this step, you end up with a 2, which is the first (leftmost) digit of the final answer.

3. **NT Calculation** time. Multiply that 2 by the leftmost digit of the divisor (keep the whole product: **06**) and the right digit of the divisor (keep only the left digit: the 0 of **08**). Add these two partial outcomes together (06 + 0 = 6).

4. Subtract this 6 from your latest **Partial Dividend** (7) to arrive at your first **Working Figure**, 1. Since **Working Figures** must be two digits, drop the next untouched digit from the dividend (4) down next to the 1 to arrive at a **Working Figure** of 14.

Review these steps a few times if you need to. Once you're comfortable with all four, move on.

You're not done with that 2 (the first digit of the answer) quite yet. Multiply this 2 by the rightmost digit of the dividend (the 4 of 3**4**). This time, keep only the rightmost digit of the product (2 × 4 = 0**8**). This is called the **U Calculation** (as in "**u**nits place," another term for "ones place"). Subtract this number (8) from the **Working Figure** (14) to arrive at the next **Partial Dividend** (6):

7,480 ÷ 34 = 2	Problem
14 (-8)	Working Figures
6	Partial Dividends

Note that the action went down here. This is the fifth and final step. Now you simply repeat this cycle.

You have a new **Partial Dividend**. What did you do with the last one?

Divide this new **Partial Dividend** (6) by the first digit of the divisor (the 3 in **3**4), ignoring any remainders (6 ÷ 3 = 2). This 2 is the next digit of your answer.

So far:

7,480 ÷ 34 = 22X

Use this new 2 just as before. First, multiply the 2 by the dividend (34), one digit at a time like before, using the special **NT** method:

2 × **3**4 = **06** (N), and
2 × 3**4** = **0**8 (T)
06 + 0 = 6

Subtract this number (6) from the most recent **Partial Dividend** (6), which results in 0. This is the first digit of the next **Working Figure**:

7,480 ÷ 34 = 22	Problem
0	Working Figures
6 (-6)	Partial Dividends

The second (rightmost) digit of the **Working Figure** will be the 8 that falls down from the next untouched digit of the dividend. Collectively, your new **Working Figure** is 08.

7,480 ÷ 34 = 22	Problem
0<u>8</u>	Working Figures
	Partial Dividends

Then multiply the new 2 (from the answer) by the rightmost digit of the divisor, using the special **U Calculation** method (keeping only the rightmost digit of the product): 2 × 4 = 0<u>8</u>.

08 (**Working Figure**) – 8 (**U Calculation**) = 0, a new **Partial Dividend**:

7,480 ÷ 34 =	Problem
08 (-8)	Working Figures
<u>0</u>	Partial Dividends

You completed another cycle; now repeat the cycle one last time. The 3 from <u>3</u>4 goes into your latest **Partial Dividend** (0) 0 times, and this will be the last digit of your answer.

Final answer: 7,480 ÷ 34 = 220.

You're not quite done; you have to keep going to see if there is any remainder. Continue solving for the remainder in much the same way.

Multiply your latest answer digit (0) by the 3 and the 4 of 34. As anything multiplied by 0 results in 0, this **NT Calculation** results in 0. As you can't take a second digit from the dividend (you're already all the way over to the right), 0 is the entire next **Working Figure**. Next, multiply 0 (from the answer) by the second digit of the divisor, using the **U Calculation** method (using only the rightmost digit of the product): $0 \times 4 = 0\underline{0}$. So, 0 (**Working Figure**) – 0 (**U Calculation**) = 0; there is no remainder. If there was, it would manifest here, as the final **Partial Dividend**.

With time, you'll get comfortable with the rhythm, and though the process may seem scary at the start, this method will become quite intuitive.

Review that problem again if you need to. When you're ready, let's do another example together: $9,876 \div 31$.

Where do you begin? Try each step on your own, then read on to check yourself.

First drop the 9 (the leftmost digit of the dividend) down to the bottom row, making it the first **Partial Dividend**.

$9,873 \div 31 =$	Problem
	Working Figures
9	Partial Dividends

Next, figure out how many times 3 (the leftmost digit of the divisor) goes into this 9. 3 goes into 9 exactly 3 times, so the first (leftmost) digit of the answer will be 3.

> 9,873 ÷ 31 = <u>3</u>XX

Time for the **NT Calculation**; take that answer digit (3) and perform some calculations. Multiply it by 3 (from <u>3</u>1), keeping the whole (N) answer (<u>09</u>). Next, multiply it by 1 (from 3<u>1</u>), keeping only the left digit (T) of the answer (<u>0</u>3). Add these two partial answers (09 and 0) to arrive at 9:

> <u>3</u> × <u>3</u>1 = <u>09</u> (N), and
> <u>3</u> × 3<u>1</u> = <u>0</u>3 (T)
> 09 + 0 = 9

With the action now moving up and to the right, subtract the number you just came up with (9) from the first **Partial Dividend** (9). This results in 0, which is the first (leftmost) digit of your first **Working Figure**, to be placed in the middle row.

9,873 ÷ 31 = 3	Problem
0	Working Figures
9 (<u>-9</u>)	Partial Dividends

The other (rightmost) digit of your first **Working Figure** comes from dropping down the next untouched digit of the dividend (8). Your **Working Figure** is therefore 08.

9,873 ÷ 31 = 3	Problem
0<u>8</u>	Working Figures
	Partial Dividends

Multiply the first digit of your answer so far (3) by the rightmost digit of the divisor (the 1 of 3<u>1</u>). This is the **U Calculation**. Keep only the rightmost digit of the product (3 × 1 = 0<u>3</u>). Subtract this number (3) from the **Working Figure** (08) to arrive at the next **Partial Dividend** (5).

9,873 ÷ 31 = 3	Problem
08 (-3)	Working Figures
<u>5</u>	Partial Dividends

Restart the cycle; how many times does the first (leftmost) digit of the divisor (3) go into this 5? Since the answer is 1, this will be the next digit of the answer.

9,873 ÷ 31 = 3<u>1</u>X

Do you remember what to do next?

You need to **NT** multiply 1 × 31:

1 × <u>31</u> = <u>03</u> (N), and
1 × 3<u>1</u> = <u>01</u> (T)
03 + 0 = 3

Subtract this 3 from the latest **Partial Dividend** (5) to arrive at 2.

9,873 ÷ 31 = 31	Problem
2	Working Figures
5 (-3)	Partial Dividends

Bringing the 7 down (the next untouched digit of the dividend), this makes 27.

9,873 ÷ 31 = 31	Problem
27	Working Figures
	Partial Dividends

Time for the **U Calculation**:

1 × 1 (from 3**1**) = 0**1**.

27 (**Working Figure**) – (this new) 1 = 26 (the newest **Partial Dividend**):

9,873 ÷ 31 = 31	Problem
27 (-1)	Working Figures
26	Partial Dividends

Restart the cycle yet again. 3 goes into 26 about 8 times. This is the final digit of the answer.

$$9,873 \div 31 = 31\underline{8}.$$

Remember, you're not quite finished; it's time to figure out the remainder. Perform the **NT Calculation**:

$$\underline{8} \times \underline{31} = \underline{24} \text{ (N), and}$$
$$\underline{8} \times 3\underline{1} = \underline{08} \text{ (T)}$$
$$24 + 0 = 24$$

26 (newest **Partial Dividend**) − 24 (**NT Calculation**) = 2:

9,873 ÷ 31 = 318	Problem
2	Working Figures
26 (-24)	Partial Dividends

Bring down the 3 from the dividend to make 23:

9,873 ÷ 31 = 318	Problem
2<u>3</u>	Working Figures
	Partial Dividends

Now it's time for the **U Calculation**. 8 (latest answer digit) × 1 (from 3**1**) = 08. Keep only the U: 0**8**. Subtract this from your last **Working Figure**:

$$23 - 8 = 15$$

You see that the final **Partial Dividend** isn't a 0, but rather 15. This is the answer's remainder, so the final answer is 318 and 15/31. If you so choose, you can come up with a rough decimal answer. 31 is close to 30, and 15 is half of 30, so 318 and 15/31 translates to a bit less than 318.5 (it's actually 318.48).

When Good Problems Do Bad Things

That last example went smoothly, but you need to learn what to do when things don't go as planned.

What If I Can't Divide by the Partial Dividend? Sometimes the first **Partial Dividend** will be smaller than the first digit of the divisor. Let's look at an example: 4,120 ÷ 69. You would first drop the 4 down, making it your first **Partial Dividend**. Based on what you've learned, this would then be divided by 6. Since 6 is bigger than 4, you have a problem.

When this happens, drop *both the first and second digit* down from the dividend to create the first **Partial Dividend**. In this example, that means you drop 41 down instead of 4, and you're good to go. 6 goes into 41 about 6 times (6 × 6 = 36). Continue with the normal strategy from there. When it's time to use the next untouched digit of the dividend as the second (rightmost) digit of your first **Working Figure**, begin with the 2 (from 4,1**2**0), since you already used all of 41.

What If The Math Breaks Down? How would you handle 6,310 ÷ 29? As usual, first drop the 6 (the leftmost digit of the dividend) down, forming your first **Partial Dividend**. Then divide this 6 by 2 (the leftmost digit of your divisor). 2 goes into 6 exactly 3 times, so the first (leftmost) digit of your answer is 3. Simple stuff.

Or is it?

It's time for the **NT Calculation**. Multiply that new answer digit (3) by 2 (from the divisor, **2**9), keeping the whole (N) answer (**06**). Next, multiply that same 3 by 9 (from 2**9**), keeping only the left digit (T) of the answer (**2**7). Add 06 and 2 to arrive at 8.

Uh-oh; this won't work. If you were to continue, you'd be subtracting this 8 from the first **Partial Dividend**, 6. Since this results in a negative number, you have to take a step backward and reduce the first digit of the answer from 3 to 2.

Try the problem now. First, perform the **NT Calculation**; you multiply the 2 (instead of 3) by 2 (from **2**9), keeping the whole (N) answer (**04**). Then you multiply the 2 by 9 (from 2**9**), keeping only the left (T) digit (**1**8). Add 04 and 1 to arrive at 5, which is less than 6, so you can continue.

This example illustrates a steadfast rule; whenever subtraction at this step results in a negative number, it means that the latest answer digit is too high and must be decremented by 1.

Quick Practice

Do a few problems on your own. The answers are listed below. For now, write out each problem in its entirety as demonstrated in the lesson.

1. 5,076 ÷ 27
2. 2,178 ÷ 121
3. 5,346 ÷ 54
4. 1,452 ÷ 88
5. 2,670 ÷ 89
6. 1,045 ÷ 55
7. 5,544 ÷ 63
8. 1,191 ÷ 12

Answers:

1. 188
2. 18
3. 99
4. 16.5 (or 16 and 1/2)
5. 30
6. 19
7. 88
8. 99.25 (or 99 and 1/4)

Single-Digit Divisors

The division method you just learned also works when dividing by a single-digit divisor. The method (for a two-digit divisor) must simply be altered by removing a few small steps. Let's walk through an example together: 904 ÷ 8.

First, like before, drop the leftmost digit of the dividend (9) down to the bottom row, making it your first **Partial Dividend**. Like before, take the leftmost digit of the divisor (8, which is the entire divisor in this case), and fit it into this **Partial Dividend**. 8 goes into 9 once, so 1 is the first digit of your answer.

No difference so far.

Next, perform the **NT Calculation**. Beginning with the **N Calculation**, multiply 1 (the first digit of the answer) by the divisor, 8, and keep the whole answer (08). With no second divisor digit, there's no need to perform the **T Calculation**. The action moves up and to the right: subtract this number (8) from the last **Partial Dividend** (9), resulting in 1 as your **Working Figure**.

As with two-digit divisors, you need a two-digit **Working Figure**. The second digit is pulled down from the next digit in the dividend (0 from 9<u>0</u>4), giving you a total **Working Figure** of 10.

With a two-digit divisor, you'd next perform the **U Calculation** by multiplying the last answer digit by the second digit of the divisor. However, this divisor has no second digit, so you can skip this step. Copy that **Working Figure** (10) straight down to the bottom row. This 10 is your next **Partial Dividend**.

From here, simply repeat the cycle.

1. 8 (divisor) goes into 10 (**Partial Dividend**) once, so 1 is the next answer digit.

2. Next, 1 (new answer digit) × 8 (divisor) = 08 (**N Calculation**).
3. 10 (Partial Dividend) – 8 = 2.
4. 2 is the first (leftmost) digit of your next **Working Figure**, and 4 (dropped down from the next untouched digit of the dividend) is the second. The full **Working Figure** is 24.
5. Again skipping the next step, drop 24 down to become the next **Partial Dividend**. 8 goes into 24 exactly 3 times, making 3 the next (and final) answer digit.
6. 3 (new answer digit) × 8 (divisor) = 24 (**N Calculation**).
7. 24 (**Partial Dividend**) – 24 = 0. This is the first (leftmost) digit of the next **Working Figure**, and there is no second. The full **Working Figure** is 0, which again gets dropped down into your **Partial Dividend**. Your remainder is 0.

The answer is 113—piece of cake. Review these steps again while you work through the problem.

Try another problem, but this time, with less help: 3,312 ÷ 9. Once you think you have an answer, read on.

Since 9 doesn't fit into 3, drop 33 down from the dividend into the **Partial Dividend**. 9 goes into 33 about 3 times (first answer digit). 3 × 9 = 27, and 33 – 27 = 6. This is the first digit of the next **Working Figure**. Dropping the next untouched digit down from the dividend, your entire **Working Figure** is 61. Skip the next step (because of the one-digit divisor) and slide 61 down to become the next **Partial Dividend**.

9 goes into 61 about 6 times; this is the second answer digit. 6 × 9 = 54, and 61 – 54 = 7. Dropping down the next untouched digit of the dividend gives you 72. Skip the next step and slide 72 down to become the next **Partial Dividend**. 9 goes into 72 exactly 8 times, making 8 the

final answer digit. Since it divides evenly and there is no remainder, you're done: the answer is 368. Voila!

Do a few problems on your own. The answers are listed below. Write out each problem in its entirety as demonstrated above.

1. $3,288 \div 8$
2. $2,628 \div 3$
3. $231 \div 7$
4. $8,194 \div 4$

Answers:

1. 411
2. 876
3. 33
4. 2,048.5 (or 2,048 and 1/2)

This process will start to feel more natural with time and practice. Because there are no carried digits or remainders, it's ideal for use in **Mental Calculation**.

Quick Review

1. The **Trachtenberg System** provides a fast, simple, adaptable method that works well for division involving one- or two-digit divisors. This method is better suited for **Mental Calculation** than classical division.

Review and Development: Section 4

In *Section 4*, you conquered some of the fear commonly associated with division by exploring a few methods for handling such problems with grace and speed. **Division Shortcuts** and the **F2D Method** are great tools for handling specific types of common division problems, and **Trachtenberg's (Fast) Division Method** is invaluable for working through longer, more difficult problems. **Trachtenberg System** division can be difficult, but once you're comfortable with it, compare it to classical division; which would you rather do in your head when it's time to tackle **Mental Calculation?**

Review

1. **Division Shortcuts**: Becoming familiar with and using a pair of simple division **Shortcuts** (the **Half-and-Half** and the **Two-and-Shift**) allows you to more easily address some common division problems.

2. **The Fraction-to-Decimal (F2D) Method**: Learning the **Fraction-to-Decimal (F2D) Method** gives you a useful familiarity with the relationships between basic fractions and their respective decimal equivalents. You can leverage these relationships to solve specific types of division problems quickly and with calculator-like accuracy.

3. **The Trachtenberg System**: The **Trachtenberg System** provides a fast, simple, adaptable method that works well for division involving one- or two-digit divisors. This method is better suited for **Mental Calculation** than classical division.

Development

Spend six days developing the skills covered in *Section 4*.

Days 1 – 2

Re-read the **Division Shortcuts** and **Fraction-to-Decimal (F2D) Method** lessons in their entirety. Focus on any aspects that were particularly challenging for you the first time around.

You should be able to recite and apply the two **Division Shortcuts** discussed in the section. You should also be able to explain the nature of the **Shortcuts** and recognize which situations are appropriate for each.

Next, continue memorizing the **Fraction-to-Decimal (F2D) Method** conversions. Even if this takes more than two days, don't move on until you're quite comfortable with them.

Days 3 – 5

For the next three days, develop your ability to solve problems with 1- or 2-digit divisors using **Trachtenberg System** division. Begin by re-reading the lesson, and then spend at least an hour each day solving such division problems using the **Fast Method**. You'll find twelve problems below to get you started, and beyond that, create your own. If you make a mistake, try to figure out where you went wrong; don't move on to the next problem until you do.

1. 986 ÷ 34

2. $115 \div 46$
3. $228 \div 12$
4. $678 \div 90$
5. $1290 \div 81$
6. $8,272 \div 22$
7. $7,315 \div 55$
8. $32,490 \div 12$
9. $88,164 \div 93$
10. $3,640 \div 8$
11. $666 \div 9$
12. $133.5 \div 3$
13. $352 \div 8$

Answers:

1. 29
2. 2.5
3. 19
4. 7.53
5. 15.93
6. 376
7. 133
8. 2,707.5
9. 948
10. 455
11. 74
12. 44.5
13. 44

When making up your own problems, they won't often end cleanly, and so you can end up with ugly remainders. This is how things work in the real world; we don't live in a universe of cut-and-dry integers. Deal with them.

Day 6

On *Day 6*, you'll solve the following ten division problems twice. The first time around, try to quickly decide which division method to use in each case, but don't actually solve them quite yet. Once you're finished, go through the problems again and solve them. The answers and most appropriate methods are listed below.

Any of these problems can be solved several ways, so keep in mind that "most appropriate method" refers to the method that would theoretically require the least time and effort, assuming you're familiar with it.

1. $21{,}451 \div 429$
2. $261 \div 5$
3. $13 \div 8$
4. $644 \div 4$
5. $35 \div 6$
6. $1{,}000 \div 875$
7. $102{,}846 \div 281$
8. $101 \div 7$
9. $5{,}094 \div 849$
10. $11 \div 9$

Answers (and most appropriate methods):

1. 50.236534 (**Trachtenberg System**)
2. 52.2 (Shortcut: **The Two-and-Shift**)

3. 1.625 (**F2D Method**)
4. 161 (Shortcut: **Half-and-Half**)
5. 5.8333... (**F2D Method**)
6. 1.142857 (**F2D Method**)
7. 366 (**Trachtenberg System**)
8. 14.428571 (**F2D Method**)
9. 6 (**Trachtenberg System**)
10. 0.818181... (**F2D Method**)

Don't get upset if your chosen method isn't the one listed in the answer. For instance, problem 6 may initially seem like a candidate for **Trachtenberg's Fast Method of Division**; however, you should recognize 875 as a familiar pattern from 7/8 (7/8 = 0.875). As a result, the problem can be solved by multiplying by 8 (giving you 8,000), dividing by 1,000 (giving you 8), and dividing by 7 (8/7 = 1 and 1/7 or 1.142857).

Likewise, problem 8 could be done any number of ways, but can be solved by the well-versed using some quick reasoning and the **F2D Method**. First, 7 fits into 70 exactly 10 times, leaving you with 31 remaining. 7 goes into 31 about 4 times, so the whole-number part of your answer is 14. As 7 × 4 = 28 and 31 – 28 = 3, you're left with a remainder of 3/7, which translates to 0.428571.

Section 5: Intermediate Calculation

If you're reading this, it either means you are a disciplined, finely tuned math machine or a filthy, stinking cheater who skipped ahead to this section. Assuming the former, congratulations—compared to where you began, you're now drastically more comfortable with the core concepts behind addition, subtraction, multiplication, division, and the intricate relationships between different types of numbers. You've studied and understand the correct methods by which to approach different problem types. You're ready for anything.

Well, except hard problems.

You can quickly come up with rough answers to many practical problems using only your mind. Armed only with a pencil and the back of a napkin, you can also quickly and confidently spew forth the precise answers to these same problems. You're way ahead of the average, and are now likely the person your family, friends or coworkers turn to when the restaurant check shows up. When baseball stats are being discussed around the water cooler. When quickly calculating the interest incurred over the term of a loan.

But you still don't know how to handle large-scale math problems. What's the best way to approach 234,129 + 32,100 + 672,397? What about 5,124,871 × 98?

Believe it or not, these problems are manageable, and you'll eventually be able to solve them using only your mind. In this section, you'll learn the final challenging calculation skills you'll need before transitioning from "calculation" to "**Mental Calculation**" in *Section 6*.

You may wonder why you should bother with more advanced topics when you can already handle most of the practical math problems that might come up in real life. There are several reasons to forge ahead:

- Learning to handle more advanced problems is a great exercise for your overall math skills. Increased comfort with numbers enables you to solve many different types of problems— practical or otherwise—much more quickly.
- Advanced math problems require a high level of concentration, especially as you transition to purely **Mental Calculation**. This forces you to develop your ability to focus, even in distracting environments.
- Solving challenging math problems can help strengthen your memory.
- Being able to quickly solve complex math problems in your mind is an impressive skill that can improve how others perceive your overall intellect.
- Lastly, believe it or not, being able to solve challenging math problems without tools can sometimes be useful. You'll see how shortly.

You'll begin by learning to add or subtract several numbers at once.

Lesson 1: Adding and Subtracting Three or More Numbers

By now, you've no doubt performed mathematical feats that would have seemed incredible to you only a short time ago. Let's keep growing.

Unfortunately, sometimes you have to add three or more numbers together. Consider itemized receipts, checkbook ledgers, multiple W2 forms and the like. Imagine you're shopping at a store with a gift card, and you must figure out if purchasing three specific items will exceed the gift card balance. In the same way, you must sometimes subtract numerical items from a list; how much would that receipt have totaled without items x and y? Of course, you could use a calculator, but with time, practice, and (eventually) **Mental Calculation**, it will be more hassle to use a calculator than to solve most practical problems on your own. This lesson will address some methods for performing such addition and subtraction tasks.

Adding Three or More Numbers Together

Let's begin with the addition of three or more numbers. Our favorite genius, Jakow Trachtenberg, has a method for doing this quickly. Just like his multiplication methods never force you to carry/borrow anything larger than a 2, Trachtenberg's addition method ensures you'll never have to count higher than 11, no matter how many numbers you're adding.

You'll first learn the method using a pencil and paper, but you'll move the work into your mind soon enough. Like classical addition, vertically align (stack) all the numbers, one atop the next, with the decimal places arranged in a nice column. If there are no decimal places, line up the numbers so that the ones place of each number is arranged in a straight column. Let's look at an example: 52.3 + 9 + 88 + 6.135. This should be lined up as follows:

```
52.3
 9
88
 6.135
```

See how all four numbers are stacked with the decimal place (whether implied or explicit) as the aligning force?

Let's add the following: 5,234 + 332 + 1,238 + 36,781 + 4,329 + 6,574. You'll first focus upon the rightmost column, adding the numbers together from the top down. When the total reaches or exceeds eleven, simply subtract eleven.

```
5234
 332
```

```
1238
36781
 4329
 6574
```

Scanning the first (rightmost) column of numbers, you see 4, 2, 8, 1, 9, and 4. You'll add the list from top to bottom, like this: "four plus two is six, plus eight is fourteen; subtract eleven. That's three. Three plus one is four, plus nine is thirteen; subtract eleven. That's two. Two plus four is six."

Next, determine the number of times you "reset" (subtracted eleven). Trachtenberg called these resets **Ticks**. You may want to make a little mark (a **Tick**) next to each digit that caused an "eleven overflow." When you're finished, count the number of **Ticks** (the **Tick Count**) for the column, and write this number below the outcome.

In the example (5,234 + 332 + 1,238 + 36,781 + 4,329 + 6,574), the rightmost column (4, 2, 8, 1, 9, 4) **Subtotal** is 6, plus two **Ticks**. For now, write these two numbers underneath the problem:

```
4
2
8 ' (note the Tick)
1
9 ' (again, note the Tick)
4
___
6 < Subtotal
2 < Tick Count
```

Now that you grasp the idea of adding columns up and recording **Subtotals** and **Ticks,** let's work through an entire problem to illustrate this method: 6,519 + 8,459 + 1,249 + 3,264. Beginning with the rightmost column, perform the addition as you just learned, subtracting elevens and recording all **Ticks.** Once you're finished with the rightmost column, continue to do this for all other columns.

```
6 5 1 9
8'4 5 9'
1 2'4 9'
3 2 6'4
_____
7 2 5 9 < Subtotals
1 1 1 2 < Tick Counts
```

After collecting all these numbers, add each column's **Subtotal** to its **Tick Count,** and following an "L-shaped" path, add the *neighboring* **Tick Count.** The neighboring **Tick Count** is the **Tick Count** to the *right* of the current column.

In this example, take a look at the rightmost column, 9 (**Subtotal**) + 2 (**Tick Count**). Because you're at the far-right side of the problem, there's no neighboring **Tick Count,** so just add 9 and 2: the right column's total is 11. This means the ones place answer digit is 1, and the extra 1 must be carried. Let's see this in action:

```
6 5 1 9
8'4 5 9'
1 2'4 9'
3 2 6'4
_____
7 2 5 9 < Subtotals
```

```
1 1 1 2 < Tick Counts
_____
      1  < Answer
```

Next, you see that the tens place (second from the right) column contains 5 (**Subtotal**) + 1 (**Tick Count**). Add 2 (the neighboring **Tick Count**), then add 1 (the carried number). The second digit of the answer is 9:

```
7 2 5 9 < Subtotals
1 1 1 2 < Tick Counts
_____
   9 1  < Answer
```

Moving one column to the left, apply the same "L-shaped" pattern: 2 + 1 + 1 = 4

```
7 2 5 9 < Subtotals
1 1 1 2 < Tick Counts
_____
  4 9 1  < Answer
```

Do the same for the next column: 7 + 1 + 1 = 9:

```
7 2 5 9 < Subtotals
1 1 1 2 < Tick Counts
_____
9 4 9 1  < Answer
```

There's one more column to go: an imaginary one placed to the left of the leftmost column. Its L-shape will touch the leftmost **Tick Count**. 0 + 0 + 1 = 1

```
0 7 2 5 9 < Subtotals
0 1 1 1 2 < Tick Counts
_____
1 9 4 9 1   < Answer
```

The end result is 19,491.

This is called the **"Ticks" Method**. Now that you've learned it, let's do another example: 1,234 + 2,345 + 8,642 + 250.

First, line the numbers up in nice, neat columns.

```
1 2 3 4
2 3 4 5
8 6 4 2
  2 5 0
```

Next, add up each column, recording the **Ticks** and **Subtotals**.

```
1 2 3 4
2 3 4 5
8'6'4'2'
  2 5 0
_____
0 2 5 0 < Subtotals
1 1 1 1 < Tick Counts
```

Finally, using the L-shaped addition pattern, figure out the final answer.

```
  1 2 3 4
  2 3 4 5
  8'6'4'2'
```

```
   2 5 0
  _____
  0 2 5 0 < Subtotals
  1 1 1 1 < Tick Counts
  _____
  1 2 4 7 1
```

Answer: 12,471.

Solve a few problems on your own using the **Trachtenberg "Ticks" Method**:

1. 983 + 12,290 + 3,938
2. 782 + 335 + 3,124
3. 983.4 + 398 + 121 + 1,289 + 801.5

Answers:

1. 17,211
2. 4,241
3. 3,592.9

Tip: After doing this a few times, you'll realize that "subtracting eleven from a number" is the same as "dropping the tens place and subtracting one from the ones place." 12 becomes 1. 17 becomes 6. 19 becomes 8. It may sound silly and obvious, but it's much faster for most people than literally subtracting the number 11.

When you feel comfortable doing so, start recording **Ticks** with your fingers instead of writing them down next to the digits. With time and practice, you'll be able to move all these **Ticks** to your head. For now, feel free to continue writing down the **Subtotals**.

Depending on the needs of the problem, when adding three or four small (two- or three-digit) numbers, it may be quicker to use **Familiar Numbers** (learned earlier in this book), rather than Trachtenberg's **"Ticks" Method**. It all depends on your personal comfort level with each method. The **"Ticks" Method** becomes more and more valuable as problems become larger.

Checking Your Work

One of the best things about Trachtenberg's **"Ticks" Method** is that it offers a simple way to check your addition with just a few additional seconds of work. It's not covered here because it's more of a luxury than a true necessity; however, it's covered in chapter four of the Cutler/McShane book, *The Trachtenberg System of Speed Mathematics.*. The **"Ticks" Method** is simple enough that once it's mastered, your final answer is usually reasonably reliable, especially when compared to classical addition.

The lesson's not over yet, but feel absolutely free to pause here and work on this new addition skill for a day or two. Create and solve problems that involve adding three or more numbers together. Use numbers pulled from your surroundings.

Subtraction Problems Involving Three or More Numbers

Subtraction problems involving three or more numbers can be complex, so it's generally best to break them down into several smaller, more manageable problems. Jakow Trachtenberg offers no guidance on this subject, so you'll have to conquer multi-number subtraction problems with sheer logical fortitude and your newfound comfort with numbers.

When a subtraction problem involves more than two numbers, it follows that two or more numbers are being subtracted from a single (usually larger) number (a **Master Value**). All the non-**Master Value** numbers should first be added together to create a **Subtraction Subtotal**. This **Subtraction Subtotal** is then subtracted from the **Master Value**.

For example: 1,421 − 399 − 211 − 49 is really just...

1,421 − (399 + 211 + 49)

which is really just...

1,421 − 659

and...

1,421 − 659 = 762

Answer: 762.

In this example, 1,421 is the **Master Value**, and 659 is the **Subtraction Subtotal**.

This system works for both rough and precise answers. If rounding, 1,421 – 399 – 211 – 49 can become 1,400 – 400 – 200 – 50, which gives you an answer of "around 750."

Let's stick with precise answers for now. Imagine you're going shopping. On your way out the door, a friend asks you to pick up some items for him (shoes and socks) and promises to reimburse you. Look at the following receipt:

Hat: $8.99
Shoes: $59.99
Jacket: $99.99
Socks: $4.99
Total: $173.96

There are two ways to deduce how much money you spent on your own purchases. The first would be to add your own purchases together (the hat and the jacket); but that wouldn't be good practice for this lesson, would it? Let's solve this problem using the subtraction method described above.

First find the **Subtraction Subtotal**, which is $59.99 + $4.99 (the prices of your friend's items). Using **Familiar Numbers** like you learned earlier, add $60 and $5 ($65) and account for the **Offset** (-$0.02) to arrive at $64.98. Then subtract this **Subtraction Subtotal** from the **Master Value**, $173.96 (the total cost). $173.96 – $64.98, converted to **Familiar Numbers**, becomes $174 – $65 ($109). Converting $173.96 to $174 leaves you with an **Offset** of $0.04, and converting $64.98 to $65 produces an **Offset** of $0.02. Remember what you learned in Section 1, *Fundamental Mathematical Concepts*: the **Final Offset** in subtraction problems is derived from the difference between the two **Offsets**. This means that the **Final Offset** is −$0.02.

This **Final Offset** is always subtracted from the pre-**Offset** answer, giving you a total of $108.98 spent on your own purchases.

Your friend owes you the rest.

Let's do another example together: 682 − 120 − 64.

Begin by adding the two smaller numbers, 120 and 64, for a **Subtraction Subtotal** of 184 (120 is already a **Familiar Number** so you can simply add them together). Subtract this from the **Master Value**, 682 (the larger number). 682 − 184 can be converted into **Familiar Numbers**: 680 − 180 = 500, minus a **Final Offset** of 2. The answer is 498.

Do the following problems on your own:

1. 288 − 15 − 6 − 74 − 3
2. 1,205 − 650 − 88
3. 90 − 1 − 44 − 14
4. 893 − 238 − 122 − 201
5. 6,810 − 782 − 2,890

Answers:

1. 190
2. 467
3. 31
4. 332

5. 3,138

Quick Review

1. When adding three or more numbers, first determine the problem's difficulty level in relation to your comfort with **Familiar Numbers**. If it's too difficult to solve using **Familiar Numbers** alone, employ Trachtenberg's **"Ticks" Method**.

2. When subtracting two or more numbers from another larger number, first add together all numbers to be subtracted (the **Subtraction Subtotal**) before subtracting them from the larger number (the **Master Value**).

Lesson 2: Multiplying Large Numbers by Two-Digit Numbers

In most cases, the multiplication skills you've already developed will be the most applicable to daily life. With shortcuts, familiarity with the multiplication tables, the wonders of the **Trachtenberg System**, and some dedicated practice, you can tear through most multiplication problems you see. You have the tools to say either:

"I went through the turnpike toll booth nine times last month at $2.15 per toll, so that's about $19." (rough)

...or...

"I went through the turnpike toll booth nine times last month at $2.15 per toll, so...5...3...9...1...it cost me a total of $19.35." (precise)

Although less common, multiplication problems involving larger numbers do come up. Being able to precisely solve these problems in your mind is a useful skill that will lessen your reliance on external tools. Learning to solve larger multiplication problems will also improve your overall comfort with numbers and their relationships, boosting your confidence and speed when performing all types of practical calculation.

To perform these feats using only your mind, you must first master the skills themselves.

We're going to enlist the teachings of our old pal, Jakow Trachtenberg. Trachtenberg teaches two methods for "speed multiplication:" the **Direct Method** and the **Two-Finger Method**. The **Direct Method** doesn't scale well, and is only marginally easier to learn than the **Two-Finger Method**. Therefore, you'll learn only the **Two-Finger Method**. Unlike Trachtenberg's unique techniques for multiplying numbers by five through twelve—which are essentially shortcuts—the **Two-Finger Method** can be used to solve any lengthy multiplication problem.

Though the **Two-Finger Method** is applicable to problems of any size, you'll begin by learning how it can be used to solve problems where one of the involved numbers is two digits in length.

The Pair-Product

First you must learn a new concept, wherein special digits called **Pair Products** are extracted from the multiplication of two numbers hidden within the multiplier and multiplicand. This is best explained with an example. Let's use 29 × 6.

Beginning from the left, if you multiply 2 (the first digit of **29**) × 6, you get 12. If you multiply 9 (the second digit of **29**) × 6, you get 54. Holding 12 and 54 in your mind, take the *second digit of the first product* (the 2 from 1**2**) and the *first digit from the second product* (the 5 from **5**4), and *add these numbers* (2 + 5 = 7). The resulting number (here, 7) is known as the **Pair-Product**. This is the core of speed multiplication, so spend time developing this before moving on.

The first two calculations you did so far (2 × 6 and 9 × 6) both produced two-digit products (12 and 54). However, sometimes one or both of these calculations will produce a single-digit number. When this happens, simply add a 0 to the beginning of the single-digit number, making it two digits.

Let's look at 83 × 3. To arrive at a **Pair-Product**, the first computation would be 8 × 3, which equals 24, from which you keep the 4. The second computation is 3 × 3, which equals 9. This is a single-digit number. Simply add a zero to the beginning (9 becomes **0**9), so you can use the first digit (in this case, 0) from this number. Then you can continue and add the two resulting values as in the previous example (4 + 0), arriving at a **Pair-Product** of 4.

Let's stop here and practice quickly figuring out **Pair-Products**. This is a strange new skill that must be carried out with speed and precision. Solve the problems, write down the solutions, and then check them against the correct answers below. If you made a mistake, try to figure out where you went wrong:

1. 18 × 4
2. 22 × 6
3. 81 × 4
4. 67 × 2

5. 16×7

Answers:

1. $(0\underline{4} + \underline{3}2) = 7$
2. $(1\underline{2} + \underline{1}2) = 3$
3. $(3\underline{2} + \underline{0}4) = 2$
4. $(1\underline{2} + \underline{1}4) = 3$
5. $(0\underline{7} + \underline{4}2) = 11$

With practice, you'll become so comfortable with this that you can start taking shortcuts. When first learning how to compute **Pair-Products**, 83×3 probably has you saying to yourself, "eight times three is twenty-four, take the four. Three times three is (zero) nine, take the zero. Four plus zero equals four." You'll soon get to the point where you say only, "four, zero, four," referring to the second digit of the first product, the first digit of the second product, and their sum, the final **Pair-Product**.

Practice finding a few more **Pair-Products**.

1. 12×2
2. 15×9
3. 92×4
4. 83×9

Answers:

1. $(0\underline{2} + \underline{0}4) = 2$
2. $(0\underline{9} + \underline{4}5) = 13$
3. $(3\underline{6} + \underline{0}8) = 6$

4. $(7\underline{2} + \underline{2}7) = 4$

The **Pair-Product** concept is the basis for Trachtenberg's **Two-Finger Method**, which is an excellent tool for solving large multiplication problems with precision. Now that you understand the basic idea of the **Pair-Product**, let's see how it's applied.

How to Multiply Using the Two-Finger Method

You'll learn the full **Two-Finger Method** by walking through some examples. The first example involves multiplying a somewhat daunting number by a two-digit number: 28,317 × 35.

First, surround the larger number with imaginary zeroes (one for each digit of the multiplier). Since the multiplier (the smaller number) is *two* digits, place *two* zeros before and after the longer number.

28,317 × 35

...becomes...

<u>00</u>28,317<u>00</u> × 35

If the multiplier was six digits in length, you'd place six zeroes before and after the longer number.

There are some new terms you should learn before continuing. Looking at the multiplier (35), think of the 3 as the **Inside Digit**, as it's closer to the multiplicand (the larger number), and 5 as the **Outside Digit**, as it's farther from the multiplicand.

Begin with the **Inside Digit** (the 3 of the 3̲5), and "connect it" to the two rightmost digits of the multiplicand (the two imaginary 0s on the right). These connected numbers are called the **Inside Pair**. Imagine connecting them with rainbow-like arches.

> Inside Pair: 0028,317**0̲0̲** × 3̲5

Next, connect the **Outside Digit** (the 5 of 35̲) to two digits in the multiplicand. To determine which two digits to use, simply shift your focus one digit to the left of where your **Inside Pair** began. Since the **Inside Pair** in this problem involved 0 and 0 (from 0028,317**0̲0̲**), you'll use 7 and its neighbor 0 (from 0028,31**7̲0̲**0). Think of this pairing between the **Outside Digit** (5) and 70 as the **Outside Pair**.

> Outside Pair: 0028,31**7̲0̲**0 × 3**5̲**

Look at the **Inside** and **Outside Pairs** again and make sure you understand where they came from before moving on. You'll start with these two connections to find your **Pair-Products**.

Now you'll apply the **Pair-Product** method you just learned to these pairs. In the **Outside Pair**, $5 \times 7 = 3\underline{5}$ (keep only the 5) and $5 \times 0 = \underline{0}5$ (keep only the 0). Added together, this gives you a **Pair-Product** of 5 ($5 + 0 = 5$). Hold this number in your mind.

Now do the same with the **Inside Pair**. $3 \times 0 = 0\underline{0}$ (keep the 0) and $3 \times 0 = \underline{0}0$ (keep the 0). Added together, this gives you a **Pair-Product** of 0 ($0 + 0 = 0$). Add the **Pair Product** you got from the **Outside Pair** (5) and the **Pair Product** you now have from the **Inside**

Pair (0); this gives you 5, which is the first (ones place) digit of the answer.

28,317 × 35 = xxxxx5

That was intense, so re-read the preceding paragraphs as many times as you like. It's a foreign concept, so it seems like more steps than it really is. Continue once you're confident in how you arrived at the first answer digit.

> *Tip: When you're still getting the hang of this method, it can be helpful to hold one finger under the number you're working with from the multiplier (smaller number), and under both numbers in the multiplicand (larger number). You'll eventually be able to eliminate this action and imagine numbers becoming "highlighted" while you're working with them. For now, it's fine to use your fingers. This is why Trachtenberg referred to this as the **Two Finger Method**.*

Moving your focus within the multiplicand one digit to the left, you see the following:

Outside Pair: 0028,3<u>1</u>7<u>0</u>0 × 3<u>5</u>
Inside Pair: 0028,31<u>7</u><u>0</u>0 × <u>3</u>5

As in the last step, address the **Outside Pair** first: 5 × 1 = 0<u>5</u> (keep the 5) and 5 × 7 = <u>3</u>5 (keep the 3). 5 + 3 = 8. Hold onto that 8. Next, do the same with the **Inside Pair**: 3 × 7 = 2<u>1</u> (keep the 1) and 3 × 0 = <u>0</u>0 (keep the 0). 1 + 0 = 1. Add the two **Pair-Products**: 8 + 1 = 9. The second (tens place) answer digit is 9.

$$28{,}317 \times 35 = \text{xxxx}95$$

The second time through probably went more quickly and made more sense. Moving on, again shift your focus one place to the left and repeat the process.

Outside Pair: 0028,**31**700 × 35
Inside Pair: 0028,31**70**0 × 35
Outside Pair: 5 × 3 = 1**5** (keep the 5), and 5 × 1 = **0**5 (keep the 0). So, 5 + 0 = 5
Inside Pair: 3 × 1 = 0**3** (keep the 3), and 3 × 7 = **2**1 (keep the 2). So, 3 + 2 = 5
Inside Pair (5) + Outside Pair (5) = 10.

This means that 0 is the next answer digit and you have to carry the 1. This carried 1 will be added to the next answer digit, so hang onto it throughout the next step of the calculation.

In case it isn't obvious, you'll continue repeating these steps until you run out of usable digits. With this example, this is when the **Inside Digit** (the 3 of the **3**5) is being computed against the 0 and the 2 at the far left side of the multiplicand (**0028**,31700). After this step, you'll only run into (preceding imaginary) zeroes in the multiplicand. You could keep going, but placing zeroes before (in other words, to the left of) a number does not affect its value; for example, 121 is the same as 000121.

Let's finish this problem together.

Our answer so far is 28,317 × 35 = xxx095.

Outside Pair: 00**28**,31700 × 3**5**

282

> **Inside Pair**: 0028,<u>31</u>700 × <u>35</u>
> 1 + 9 (+ the carried 1) = 11. Use the ones-place 1; carry the tens-place 1
> Answer so far: 28,317 × 35 = xx1,095

then...

> **Outside Pair**: 00<u>28</u>,31700 × 3<u>5</u>
> **Inside Pair**: 0028,<u>31</u>700 × 35
> 4 + 4 (+ the carried 1) = 9
> Answer so far: 28,317 × 35 = x91,095

then...

> **Outside Pair**: 0<u>0</u>28,31700 × <u>3</u>5
> **Inside Pair**: 00<u>28</u>,31700 × <u>35</u>
> 1 + 8 = 9
> Answer so far: 28,317 × 35 = 991,095

then...

> **Outside Pair**: <u>00</u>28,31700 × 35
> **Inside Pair**: 0<u>0</u>28,31700 × <u>3</u>5
> 0 + 0 = 0

At this point, you're out of usable digits (other than zeroes), so the answer is 991,095. There are no secrets or exceptions; the method is relatively straightforward when all is said and done. Let's do one more problem together before you try a few on your own.

351 × 23.

Think about this for a second. How will you begin?

351 × 23

becomes...

0035100 × 23

Isolate and begin working on the first set of **Inside** and **Outside Pairs**. You can calculate the **Inside Pair** before the **Outside Pair** or vice-versa; the order doesn't matter. To demonstrate, this time we'll switch up the order in which you approach the two. When you're on your own, you can solve them in the order you prefer.

Inside Pair: 00351<u>00</u> × <u>2</u>3
Outside Pair: 0035<u>1</u>00 × 23<u>3</u>
0 + 3 = 3
Answer so far: 3

then...

Inside Pair: 0035<u>100</u> × <u>2</u>3
Outside Pair: 003<u>5</u>100 × 2<u>3</u>
2 + 5 = 7
Answer so far: 73

then...

Inside Pair: 003<u>51</u>00 × <u>2</u>3
Outside Pair: 00<u>35</u>100 × 2<u>3</u>
0 + 10 = 0; carry the 1
Answer so far: 073

...then...

That's it; the answer is 8,073.

Perform the following calculations; the answers are below. Don't cheat yourself; figure out and write down the answers, then check them. Only write down the answer digits, never the problem steps or work. If you've made a mistake, try to figure out where you went wrong.

1. 39,281 × 21
2. 1,382,938 × 40
3. 23,892 × 37
4. 19,230 × 22
5. 982,973 × 89

Answers:

1. 824,901
2. 55,317,520
3. 884,004
4. 423,060
5. 87,484,597

Take a moment to appreciate the fact that Trachtenberg figured all of this out in his head while staring at a wall in a prison camp. These methods are evidence of truly humbling genius.

Feel free to stop here and spend a day or two working on this method. Aggressively practice multiplying numbers, as long as one of the numbers is exactly two digits in length and the other is fairly large.

While practicing the **Two-Finger Method**, continue to review the multiplication methods you learned in previous sections. Later, in *Advanced Calculation*, you'll scale this method so you can apply it to problems of any size. Beginning with two-digit multipliers gives you the basic tools necessary to face more daunting multiplication problems down the road.

Quick Review

1. The **Trachtenberg System** provides a method for multiplying two-digit numbers by numbers of any size.

Review and Development: Section 5

Though short, *Section 5* covered some pretty complex topics, and it will take time to implement it all with any grace.

As mentioned, while you're developing these skills, you should continue working to improve upon all the other skills you've explored so far in *Master the Language of the Universe*. As you learn more, neglecting to practice the fundamentals could cause you to regress in these abilities. This, in turn, could make you feel disheartened, setting in motion a cycle of unlearning and negativity that could undo all your hard work. Stay current, stay hungry, and enjoy the metamorphosis.

Review

1. **Adding Three or More Numbers**: When adding three or more numbers, first determine the problem's difficulty level in relation to your comfort with **Familiar Numbers**. If it's too difficult to solve using **Familiar Numbers** alone, employ Trachtenberg's **"Ticks" Method**.

2. **Subtracting Three or More Numbers**: When subtracting two or more numbers from another larger number, first add together all numbers to be subtracted (the **Subtraction Subtotal**) before subtracting them from the larger number (the **Master Value**).

3. **Multiplying Large Numbers by Two-Digit Numbers**: The **Trachtenberg System** provides a method for multiplying two-digit numbers by numbers of any size.

Development

Spend five days working on the skills learned in *Section 5*.

To continue moving ahead, you need to be able to perform these types of calculations without too much dedicated thought. Next, you'll start learning to perform these methods using only your mind.

Days 1 – 2

Spend time on *Day 1* reviewing everything you learned in *Section 5*. Take extra time with any aspects that you found particularly challenging. Use the remainder of this day and the next to work on your addition and subtraction skills for problems involving three or more numbers. Begin by arbitrarily selecting four or five numbers (between three and six digits long), adding them together, and checking your answer with a calculator. Then choose four numbers and subtract the smaller three numbers from the largest (the largest number should be significantly bigger than the other three to avoid getting into negative numbers). Begin to enjoy the process and the challenge of solving these problems, and it will become less of a chore than you'd imagine.

Days 3 – 5

On *Day 3*, briefly re-read the chapter on **Trachtenberg System** multiplication involving two-digit numbers. Spend three days practicing this skill on your own. Simply choose two numbers (one two digits long, the other four or more), multiply them, and check your answer with a calculator.

Section 6: Doing It All in Your Head

At this point, I believe I've fulfilled my mission of imparting *practical* math skills. Pretty much everything from here on out can be considered bonus material.

What is Mental Calculation?

As the name implies, **Mental Calculation** refers to the act of solving arithmetic problems in your head, without the aid of calculators, your fingers, or any other tools, including analog tools like a pencil and paper. The term also implies an expectation about the speed with which one performs calculations, as even the untrained can usually crank out a solution given enough time.

Why is Mental Calculation Useful?

The way you were taught to do math in school is standardized, easy to remember, and perfect for developing initial math skills; however, it doesn't effectively leverage the way your brain works, and thus isn't efficient for unassisted calculation. For example, in classical math, carried numbers and remainders must be tracked and remembered while continuing to perform calculations in the forefront of your mind. This isn't complicated with a pencil and paper, but it's error-prone when hashing out calculations in your head. Many of the methods you just learned were selected because they're superior to classical math when it comes to **Mental Calculation**.

Even someone who doesn't handle numbers professionally encounters arithmetic on a daily basis: creating invoices, paying bills, calculating sales tax, and comparing prices. Many people are involved in fantasy sports leagues, clip coupons, or bet on horse races. Other people grade history papers, sign time sheets, or sell vintage Christmas tree ornaments on eBay. That arithmetic isn't always easy, but just because a math problem is complex doesn't mean that you shouldn't be able to solve it in your head.

It's not easy, and it's certainly not for everyone, but it *can* be done.

How We Are Going to Approach Mental Calculation

Mental Calculation consists of three components. They are:

- Ability
- Memory
- Focus/Concentration

Ability

Ability refers to familiarity with the calculation methods you need to actually solve a given problem. By now, ability should be well taken care of, as most of this book so far has been dedicated to acquiring these tools. You've developed all the skills needed to solve addition, subtraction, multiplication, and division problems at various levels of complexity. You have mental flowcharts that help you to determine the most appropriate methods for most problems you may face. If you're still uncomfortable with any skills you've learned, take the time to develop them before moving on.

Memory

Memory is the ability to temporarily store information required for a problem you're solving. When solving a problem using **Mental Calculation**, memory is needed for:

1. Remembering the problem;
2. Remembering your chosen solution method (and how to perform that method);
3. Remembering which solution step you're on; and
4. Remembering the digits of the answer as you begin to uncover them.

That's quite a bit of memorization, which is especially challenging when you need to memorize swiftly. This can become more manageable with well-directed practice.

Problems involving one- or two-digit numbers can usually be solved using your natural working memory, while more complex problems may require mnemonic devices to remember the necessary information; by the same token, you must also be able to instantly convert these **Mnemonic Clues** back into usable figures.

We'll go through a basic **Mnemonics** primer before moving on, but keep in mind you'll only need to use these tools for more complex problems.

Focus/Concentration

What's the hardest part of doing a math problem in your head? For most people, it's "digit stability"; while you're working, the numbers have a tendency to change or decay in your mind. Even when they don't, you often fail to trust yourself to hang on to numbers reliably. This is where focus/concentration comes in.

Focus/concentration refers to the ability to stay completely engaged with a problem (and its solution) amidst distractions. Assuming you're not solving math problems in an isolation tank or at the edge of a serene lake, you need to be able to perform under the pressures posed by external stimulation and perceived time restrictions. Concentration can be a problem even when using external aids, and once the external aids are gone, the need to focus increases exponentially. During every single step of a **Mental Calculation** problem, you'll be completely reliant upon your ability to concentrate effectively.

You'll develop your ability to perform feats of **Mental Calculation** in two parts, using a proprietary method called the **5-7 Approach**:

- You will advance in *five* phases of incrementally increasing complexity, easing your way from pencil-and-paper work to end-to-end unassisted **Mental Calculation**.

- Each of the five phases incorporates different combinations of *seven* distinct skills.

The rest of this section will familiarize you with these phases and skills.

Lesson 1: Single-Digit Mnemonic Clues

When memorizing longer numbers, you're going to need to use tools beyond your natural working memory. Numbers are mundane pieces of data, and the trick to memorizing information like this is to associate it with visual, imaginative scenes and scenarios. Here, you'll learn to store a single-digit number in a specific "space" in your memory by leveraging your imagination.

There are two concepts key to memorizing small numbers: the **Mnemonic Vocabulary** and the **Journey Method**. Together, they form a tight bond that allows for incredible feats of memory to take place.

The Mnemonic Vocabulary

A **Mnemonic Vocabulary** is a series of words, ideas, or objects used to memorize a specific type of information. For example, if you were to use the color red to remember January, blue to remember February, yellow to remember March, and so on, these colors together form a **Mnemonic Vocabulary** that represents the months of the year.

According to everyone from the ancient Greeks to modern competitive mnemonists (memory competitors), the key to memorizing small numbers is to create **Mnemonic Clues** that represent each number. These **Clues** will serve as your personal **Mnemonic Vocabulary** for single-digit numbers.

Let's create a basic set of **Single-Digit Mnemonic Clues** for all ten single-digit numbers (0 through 9).

Take the number 1, for example. Two examples of possible visual **Clues** for the number 1 are "gun" (which *sounds* like "one") and "pencil" (which is long and thin, so it *looks* like the written character "1"). Examples for 9 could be "wine" (which *sounds* like "nine") and "lollypop" (which sort of *looks* like a "9").

How about 8?

What objects or images look similar to the number 8? One doughnut sitting atop another? A snowman? John Lennon-esque sunglasses? A bra hanging on a laundry line? What about a few objects or images that *sound* like "eight?" A dumbbell (weight)? A gate? Get as creative as you like. How about a kangaroo ("g'day, *mate*")? Choose a single image to use as your **Single-Digit Mnemonic Clue** for the number 8.

The most important thing is that the **Clue** you choose makes sense to *you* and will be easy for you to instantly recall/translate. Other people's suggestions—though they may give you ideas—ultimately don't matter.

You don't necessarily have to choose a **Clue** that looks or sounds like the number; your first instinct is often the best choice: it represents a deeply-rooted gateway into your subconscious, and is therefore likely to be the easiest association to recall quickly. So hey, if you think the number seven looks like a green baseball bat wearing a cowboy hat, so be it. Don't shy away from absurdity, inappropriateness, or downright perversion, as long as it makes *immediate sense to you*.

Do this for every digit from 0 to 9. Grab a pen and fill the table below with your **Single-Digit Mnemonic Clue** images. Choose a single **Clue** for each digit. If you get stuck, read on.

Number	Your Single-Digit Mnemonic Clue
0	
1	
2	
3	
4	
5	
6	
7	
8	

As I said, you should really come up with your own **Single-Digit Mnemonic Clues**. However, if you're stuck, you'll find a few suggestions below. When you're first starting out, it can be difficult to think with creative freedom, but I promise you it will come with time.

#	Looks Like	Sounds Like
0	Baseball, soccer ball, bowling ball, golf ball, basketball, eyeball, apple, clock, Frisbee, bowl, egg, peach	Hero, gyro, Nero
1	Pen, pencil, marker, feather, baseball bat	Nun, gun, sun, bun, clown (fun), sneaker (run)
2	Goose, sickle	Shoe, lawyer (sue), gnu
3	A heavyset person's profile, heart, breasts, buttocks	*Spree* (the candy), flea, knee, bee, tree
4	Knife, triangle, pizza slice	Boar, gore, bore, lore, door, whore, someone's back turned to you (ignore), celebrating soccer player (score)
5	Wheelchair, brimmed hat, snake	Chive, dive, jive, hive, Frankenstein's monster ("it's alive!"), car (drive)
6	Teapot, a stick figure doing a handstand, apple with a big stem	Hicks, sticks, candles (wicks), bricks, chicks, a thirsty dog (licks)

7	Arrow, ledge, elbow, uni-brow, hook, candy cane	Viking hat (Sven), rising bread (leaven), cloud (heaven)
8	One doughnut sitting atop another, snowman, bra, John Lennon-esque sunglasses	Weight, gate, angry face (hate), bait
9	Chain mace, Pac-Man	Twine, vine, mine, wine, sign, line

We think in pictures and remember in an elaborate and free-flowing amalgamation of emotions and visuals. To harness this is to tap into the mind's natural memory mechanisms.

Your **Clues** should be flexible and dynamic; as you'll soon see, they will be put to use in a wide range of contexts. For example, if you're using "duck" as a clue, the imaginative scenes you'll need to create might require a tiny, pocket-sized duck; a city-sized duck; a red duck; or an anthropomorphic, humanoid duck in a suit and tie. Your imagination is boundless and answers to no one but you. Eggs can be blue and hatch baby cars. Trumpets can shoot lasers. Sharks can print receipts. The moon can fart confetti. Let yourself go and get weird.

Once you've established your **Single-Digit Mnemonic Clues** for the digits 0 through 9, take some time and commit them to memory, as you'll need them shortly. Don't move on until you can recite the ten digits and their respective **Single-Digit Mnemonic Clues** with relative ease.

The Journey Method

Next, we'll turn to the **Journey Method** of memorization. According to many scholars, the ancient Greeks first developed this system thousands of years ago, but its roots could possibly be even older. The **Journey Method** is still used by almost every noteworthy modern mnemonist.

The core idea is simple: first, select an ordered pathway along a familiar course (for example, walking through the rooms of a friend's home or through a familiar park). Then, with this pathway in mind, imagine placing **Mnemonic Clues** (for example, the **Single-Digit Mnemonic Clues** you just memorized for the numbers 0 through 9) at each location encountered along the path. These individual locations are simply called "**Steps**," and the whole pathway is called a "**Journey**."

Recent scientific research has revealed that a part of the brain called the hippocampus houses specialized cells that are used specifically to keep us oriented in space. These cells (called "grid," "place" and "border cells") work with the memory to provide accurate spatial records of the places we've been, including their respective layouts, features and scales. The **Journey Method** leverages this function.

Building Your First Journey

Let's begin with a basic exercise to get comfortable with the core concepts of the **Journey Method**. We're going to build a **Journey** based on wherever you're sitting right now. This **Journey** is going to consist of three **Steps**.

The first is where you're sitting right now; it's the table in front of you, your lap, or your chair. Adapt it based on your specific situation.

The second **Step** is the floor nearby, halfway between your current location and the nearest door or entryway. Look at this spot for a moment, and truly take it in; try to notice minute details you might otherwise overlook.

The third **Step** is the nearest door or entryway; take a moment to study this **Step** as well.

That's three **Steps**: your current location, the floor halfway between you and the nearest door, and the nearest door.

Let's say you want to memorize the number 673. First, imagine a **Clue** for 6 (what was your **Single-Digit Mnemonic Clue** for 6?) in your current location (the first **Step**). Then, imagine a **Clue** for 7 on the floor (your second **Step**), and a **Clue** for 3 at the nearest door (your third and final **Step**).

Shut your eyes and picture the **Clues** in their respective locations. Don't simply think about the objects sitting there; rather, work them into the scene in imaginative ways. Push the boundaries of your creativity.

Review the three **Clues** and their respective **Steps** several times. Once you feel comfortable with the **Journey** you've built, walk through the three **Steps** and "retrieve" the **Clues**, translating them back into digits as you go.

This exercise introduces the basic concepts behind the **Journey Method**: things we need to remember (or **Clues** about these things) are placed at each of the individual **Steps** of an imagined **Journey**.

Building a Longer Journey

Take some time and build your own **Journey** of ten **Steps**. For simplicity's sake, start with a familiar place, such as your home.

Make sure to follow logical, natural paths. Don't "teleport" from an upstairs bedroom directly to your basement. If you can't get from *point A* to *point B* naturally in real life, the path won't seem natural to you when you're working through it in your imagination.

A good rule of thumb is to walk through the home as though you're giving someone a tour. Think of how a real estate agent would guide a prospective buyer.

Say each **Step** aloud. "**Step** one is my front door. **Step** two is my living room..." and so on, until you reach the tenth **Step**. Say them again and again out loud—a bit faster each time—while imagining not only each **Step**, but also the trip or transition from one **Step** to the next.

Take a number that would ordinarily be difficult to memorize quickly—2,184,094—and memorize it using the **Journey** you just developed. Unlike the first exercise, the starting point of this **Journey**—though familiar—won't necessarily be directly in front of you; you'll have to rely entirely on your memory of the layout.

Imagine the following in great detail: you're alone at the first **Step** of your **Journey**. Take your time and get lost in the scene. What's the temperature? What are you wearing? Is it day or night? In your fantasy, imagine that you shift slightly, and your foot touches something. You look down to find a shoe (shoe sounds like "two"). Imagine your reaction; why is there a shoe on the ground? How would you react in the real world? What does the shoe look like? What does it feel like? Is it a men's or women's shoe? Formal footwear or sneaker?

A shoe may not be your chosen **Single-Digit Mnemonic Clue** for the number two; I'm using **Single-Digit Mnemonic Clues** from the examples only as a guide to help get you used to the process. You should substitute your own **Clues** as needed.

Next, you arrive at the second **Step**, and see that there is a gun lying on the floor ("gun" sounds like "one"). Again, make the image vivid and realistic. Picture the gun; pick it up, feel it, and notice subtle details about it. Is it cold? Is it heavy? Remember that you can be as creative as you want. Instead of a gun, perhaps 1 could be a tiny clown ("fun")? What would a tiny clown look like? Feel like? What would it do?

Now, wander toward your third **Step**. Where are you? Picture as many details as you can. Wherever you find yourself, imagine a woman's bra hanging on the wall, as though it was a piece of art (a hanging bra looks like the number "eight"). What color is it? Is it frilly or plain?

You get the idea: in order to memorize this seemingly complex number, use **Steps** along a familiar path (**Journey**), leaving for yourself at each one a sole **Single-Digit Mnemonic Clue**. Continue memorizing the number 2,184,094 using your ordered **Steps**. As you're using a ten-**Step Journey** to memorize seven **Steps**, you'll end up with three left over.

Take a deep breath, relax, shut your eyes, and imagine yourself walking through your **Journey** from the very beginning. Along the path, retrace your strange experiences, recollecting and deciphering into numbers the **Single-Digit Mnemonic Clues** you uncover. The imaginary events that play out within a **Journey's Step** are referred to as a **Scene**. This method creates such strong impressions that you can usually remember the **Clues** for a long time (especially if you create quirky, distinct scenarios at each **Step**).

> *Tip: Once you become comfortable with a particular **Journey**, you may find yourself imagining the transitions between the **Steps** occurring quickly, as though in "fast forward." This is perfectly natural.*

Go through your **Journey** once again and retrieve the stored numbers; this time, try to go a bit faster.

The above is an excerpt from my book, *Never Forget Again: Master Your Imagination and Develop a Legendary Memory*; if you're interested in diving far deeper into number memorization, I strongly suggest taking a look, as **Single-Digit Mnemonic Clues** represent just the tip of the iceberg.

Don't move on until you're relatively comfortable quickly recording and holding about ten digits in your head for several minutes. This is about how many digits are required to represent a mid-sized problem and solution. If this is foreign to you, it can take quite some time to gain control of your memory to such an extent. Be patient and your efforts will pay off.

Quick Review

1. Numbers are a good example of mundane information, which your brain tends to discard. In order to retain numbers, you must associate each digit with an imaginative **Clue**, which your brain can usually retain much more easily.

2. A **Mnemonic Vocabulary** is a personalized set of **Mnemonic Clues** that represent specific pieces of mundane information.

3. Creating **Single-Digit Mnemonic Clues** (standardized images or objects that represent each number from 0 through 9) allows you to easily imagine and recall individual digits. Some of the most popular **Single-Digit Mnemonic Clues** are those that either physically resemble ("2" and "a cobra") or sound like ("three" and "tree") the digit.

4. A **Journey** is an imagined path through a familiar setting in which you store information. You can do so by integrating **Mnemonic Clues** (such as **Single-Digit Mnemonic Clues**) within the individual **Steps** that comprise the **Journey**. You can then mentally retrace the **Steps** of the **Journey**, observing the **Mnemonic Clues** and translating them back into information (such as numbers).

Lesson 2: The Art of Concentration

Here, you'll find some strategies that are useful when learning anything that requires total concentration.

How to Concentrate on <u>Nothing</u>

You're going to begin by learning to concentrate on nothing. That's right—nothing. You'll learn to lock out distractions, clear your mind, and create a chamber of isolation within your mind.

Location, Location, Location

You'll eventually be able to concentrate regardless of location, but as you're just getting started, you should isolate yourself as much as possible while practicing this skill. Begin in a place with minimal distractions; you can go into a bedroom and shut the door (refrain from laying down so as to minimize the risk of falling asleep); you can work on this while showering, or while walking through a serene area such as a park or graveyard. Believe it or not, there are still places in the world—all around you, unless you live in downtown Beijing or New York—that subject you a bit less stimulation than you're used to; you simply have to seek them out. In addition, while beginning to practice this skill, turn off all mobile devices.

Breathing

Especially when beginning, much can be learned from Vinyasa (or "Flow") Yoga. Concentrate on your breathing, and make sure to breathe in a controlled, even rhythm. Your stomach should expand outward and contract inward, with very little raising of your shoulders. If you're sitting, put your feet flat on the ground and your palms flat on a table or desk and imagine them slowly growing roots that crawl deep down into the earth. Concentrate on relaxing for a minute or so before attempting to practice concentration. Don't fall asleep! Spend a few sessions just getting this far—lock your mind in with your breath. Next, become aware of the external air pushing against your skin, and imagine erasing your skin so that the air flows in and out of your body freely.

This will always be your first step when partaking in concentration exercises.

The Black Rectangle

As you get pulled away into thoughts and distractions, you need to yank your consciousness back into the place you'd like it to be. You're going to learn to stop letting your attention rule you, and instead, learn to rule your attention. This will become more natural after time, but for now, your attention is most likely going to act like an undisciplined child.

Focus on your breath until you're able to bring yourself into the relaxed state described above. Clear your mind, shut your eyes, and attempt to imagine "nothing." Since it's virtually impossible to imagine nothing or even contemplate truly empty space, your next best option is to visualize a representation of nothingness: A black rectangle that takes up almost all of your field of vision. Imagine nothing else—no sounds, no other images, and no smells or emotions. Allow no thoughts, to-do lists, anxieties, or anything of the like to permeate this rectangle. Just imagine this simple black rectangle and, still relaxed, focus wholly upon it.

Can you do it? How long does it last before other ideas start flooding your mind? Before you get an itch on your forehead? Before you become aware of your tongue or the sounds caused by your breathing and swallowing? How long before boredom sets in, and you're whisked back into your hectic reality?

Furthering the visualization, imagine that the outer border of the black rectangle literally blocks incoming distractions; they touch the edges of the rectangle and get deflected away, out of mind. Only the rectangle remains.

Spend a few minutes several times each day getting into your breathing place, and then imagine this black rectangle of nothingness for as long as you can. With time, you'll learn to more quickly construct this visual, concentrate on it, and deflect other thoughts. Once you're comfortable enough to imagine this black rectangle—and nothing else—for about five minutes without other distractions getting in the way, begin to attempt this visualization strategy in more distraction-laden areas (on subway cars, in cafeterias, etc.).

Obviously, don't try this while driving or operating a forklift.

When you're ready, begin attempting this exercise with your eyes open, but with your focus unfixed. Despite the open eyes, try not to "see." Once you can concentrate on the black rectangle in a distracting setting, with your eyes open, and truly block out a significant amount of what is going on around you, it's time to move on. Next, you'll learn to use this black rectangle as a blank canvas upon which you'll concentrate on a specific thought or problem.

How to Concentrate on <u>Something</u>

To begin, you are going to memorize the number 4220. Within your black rectangle of distraction-resistance, imagine the numbers written in a high-contrast manner, such as large black text atop a brightly-colored background. What is the font? How big is it? What color did you decide upon for the background? This is how you focus on single objects or ideas: Focus on them literally and imagine them vividly and stably. Truly look at the number as though you are seeing something in the real world. If this seems difficult, just keep working on it. It's a strange and new skill, and development will take time and personalization. You're working out your visualization muscles.

This can also be used for **Mnemonic Scenes**: At first, imagine the scene occurring inside the black rectangle, as though it were a television. Then, imagine stepping within the scene and immerse yourself in it as a participant. Remember, you're inside the black rectangle, so you remain immune to outside distraction or influence. Nothing else should be allowed in—this setting should be occupying the entire spectrum of your concentration. Paint the scene vividly using the fullest extent of your imagination, but don't let anything else distract you. With continued, controlled breathing, further immerse yourself in the scene's details, as though it was your reality.

Once you feel that you're successfully able to do this, it's time to learn how to concentrate on several things at once.

How to Concentrate on <u>Several Things at Once</u>

The short version: You can't, so don't even try. True simultaneous attention to multiple cognitively-challenging tasks is nearly impossible for most of us.

Many years ago, I attended a management seminar in which the instructor was trying to convey the importance of uninterrupted focused time. She had three different people approach a whiteboard simultaneously and perform the following tasks:

Write out the numbers one through ten first in Arabic numerals ("1," "2..."), then spelled out ("one," "two..."), and finally as Roman numerals ("I," "II...")

Do the same thing, but focus on one number group at a time (write "1, one, I"; "2, two, II"; "3, three, III...")

Here are the speeds (in seconds) in which the three individuals were able to perform said tasks:

Task 1: 22.9 seconds; Task 2: 47 seconds

Task 1: 21.8 seconds; Task 2: 27.7 seconds

Task 1: 21.1 seconds; Task 2: 32 seconds

Note that the same end result was achieved through task 1 as was through task 2, but in every case, it took drastically longer to perform the version of the task that constantly derailed the individual's train of thought.

The takeaway from this is that the human mind is single-threaded—that is to say that you can only truly concentrate on one thing at a given time. Sure, you can walk and talk simultaneously, or dance poorly while making a grocery list, but when it comes to more taxing mental tasks, multitasking becomes exceedingly difficult to do with standard brain wiring. The trick to multitasking isn't *trying really hard to focus upon two things at once*; rather, it's recognizing your inability to properly pay attention to multiple things at once, accept it, and more intelligently distribute your attention. Comedian Dmitri Martin can famously create two totally separate drawings at the same time—one with each hand. It could be argued that on the most basic cognitive level, he isn't truly focusing on two things at once, but is rather distributing his concentration back and forth at a rapid rate.

Get yourself into a focused mental state, as you learned to do. Free yourself of distractions, achieve a good breathing rhythm, and conjure up your black rectangle. Let's do an experiment inspired by the task demonstrated in the management seminar. We're going to start with the number 1 and the letter Z. You're going to simultaneously increment the number (1, 2, 3…) and decrement the letter (Z, Y, X…) in your mind. As you learned, "simultaneously" is a bit of a misnomer, as you'll actually be sharing your attention between the two tasks, rapidly shifting your focus from one to the other.

Say each number/letter combination out loud (1/Z, 2/Y, 3/X…). Try to pair each combination as closely as possible (that is, you should be saying "1Z," not "1… …Z"). This is a purposely difficult exercise, in which you have to keep track of two independent tasks. Use the black rectangle to reduce distraction, and imagine one item (the number) on the left side of the black rectangle's interior and the other item (the letter) on the right side.

Over time, your ability to visualize stable images in your mind will improve, as will your ability to block out distractions. This is important because in order to perform calculations upon multiple numbers using only your mind, you'll need to store said numbers for a period of time.

Quick Review

1. **Focus/Concentration** refers to the ability to stay completely engaged with a problem (and its solution) amidst distractions. In order to develop this skill, you should progress through three levels of increasing difficulty: First, learn to concentrate on *nothing*; then, learn to concentrate on *something*; finally, learn to concentrate on *several things at once*.

Lesson 3: Mental Calculation—The Five Phases

"NUMBERS EXIST ONLY IN OUR MINDS. THERE IS NO PHYSICAL
ENTITY THAT IS NUMBER ONE."

—RAYMOND A. BEAUREGARD, MATHEMATICIAN

Beauregard is correct; Numbers, though used to describe objects
or ideas, are themselves abstractions. Despite their intangibility,
numbers are often manipulated like physical things, shifted around and
intermingled to create other numbers with vastly different values and
meanings. As you learn to hold and manipulate numbers in your mind,
the lack of a physical equivalent becomes increasingly apparent. You
will be forced to exert more control to keep the numbers stable and
intact. For this reason, you can't "wing it"; you need a process.

Your journey from pencil-and-paper-reliant, long-dividing
novice to crowd-wowing mental mathlete will take place over five
phases, mastered one-by-one with time and practice. Each step contains
a new challenge not found in the previous step; this new challenge is
written in *italic*.

1. Reference the problem on paper (or screen, etc.) as needed, perform your work on paper, record the answer on paper, and announce the answer (you're already at this point).

2. Reference the problem on paper as needed, *do the work entirely in your head*, record the answer on paper, and announce the answer.

3. Reference the problem on paper as needed, do the work entirely in your head, *record the answer in your head*, and announce the answer.

4. *Briefly view the problem on paper (without continuing to reference it)*, do the work entirely in your head, record the answer in your head, and announce the answer.

5. *Listen to the problem (verbally)*, do the work entirely in your head, record the answer in your head, and announce the answer. Alternately, store the answer in your mind for later retrieval.

The next three lessons will cover the seven distinct skills involved in **Mental Calculation**. Each of the five phases listed above requires one or more of these skills. By implementing these skills effectively and in a specific order, you'll be able to reach the elusive and challenging fifth and final phase.

Quick Review

1. There are five phases of **Mental Calculation**, each a bit more challenging and impressive than the last. Reaching the fifth and final phase of true, unassisted **Mental Calculation** will require all seven distinct **Mental Calculation** skills outlined in the next three lessons.

Lesson 4: Mental Calculation—The Seven Skills (Part 1)

There are seven distinct skills involved in solving an end-to-end **Mental Calculation** problem. Not all five phases described in the previous lesson require *every* skill, but each phase involves at least one of them. As the phases become more challenging, you'll need more of the skills to conquer them.

Skill #1: Storing the Problem

Since you can reference the problem as needed during phases 1, 2, and 3, storing the problem is unnecessary. The problem is assumed to be static, displayed such that you could stare at it as long as you like. During phase 4, you get to see the problem briefly, and during phase 5, you only hear the problem conveyed orally. Steps 4 and 5 both pose a temporary storage problem.

When presented with a simple problem involving digits that can be stored using your natural working memory (five total digits or less, like the problem "56 × 132"), don't bother using mnemonic tools; instead, simply remember these small numbers through brute force memorization.

However, to properly store larger problems, it's best to convert the numbers involved into a series of mnemonics. Use **Single-Digit Mnemonic Clues** and store the digits one-by-one in a **Mnemonic Scene**. This would mean seeing/hearing "nine thousand, one hundred and forty-four" should prompt you to store "9...1...4...4," one digit at a time.

Hearing a number conveyed orally presents an additional challenge over having it written out in front of you, as number-to-mnemonics translation has to be done on the fly. This can be exceptionally difficult if the number is spoken quickly. The only way to develop this skill is to practice.

Time to Practice

Let's look at a few numbers together. Memorize each of the following numbers as quickly as possible. Only memorize one at a time; the goal isn't to memorize several numbers at once, but to memorize a single number as quickly as possible using **Single-Digit Mnemonic Clues**. After memorizing the first number, look away and do something else for a few minutes while keeping that number in the forefront of your mind.

1. 328
2. 1,904
3. 2.832
4. 12,680

5. -562
6. 8,289
7. 44.66
8. -18.66
9. -9,137
10. 982,587

Once you're comfortable with memorizing written numbers, it's time to move on to spoken ones; if possible, have a partner recite arbitrarily chosen numbers (between four and eight digits) to you out loud. Memorize each number as quickly as possible.

Don't move on until you're confident in your ability to store individual numbers as described. Take as long as you need.

Skill #2: Isolating the Individual Parts of the Problem

You're now proficient in memorizing individual numbers; however, most problems involve two or more numbers that are subject to a mathematical operation, such as addition or division.

Even if the problem is small, don't store multiple numbers in a single **Journey**; it's better to use separate **Journeys** to represent each number. In almost all cases, you can find two locations in your immediate setting with which to store **Journeys**. On top of memorizing and isolating the two or more numbers involved in a problem, you must remember whether you're adding, subtracting, multiplying or dividing. As the type of mathematical operation is a relatively simple thing to remember, I encourage you to rely upon your natural working memory for this.

Let's practice.

First, store the following problems in your head one at a time. Look at one, store it as quickly as possible, wait a minute or so, then and repeat it back in three pieces; the first number, the second number, and the whole problem, including the operator. Don't worry about actually solving them.

Take 439 + 1,086 for example; you would say, "Four hundred thirty-nine. One thousand eighty-six. Four hundred thirty-nine plus one thousand eighty-six." Make sure you're using your **Journeys** to store and extract the numbers rather than repeating them from working memory. Like with the last practice session, don't worry about memorizing the problems all at once; memorize a single problem, then distract yourself with an unrelated task while keeping it fresh in your mind. There are quite a few practice problems here because this skill is vitally important to master before moving forward.

1. $28,416 - 6,188$
2. $9,825 + 3,089$
3. $2,345 - 108$
4. $483 \times 1,441$
5. $8,488 \div 720$
6. 11×865
7. $839 \div 75$
8. $1,285 - 515$
9. $3,750 \div 5,182$
10. $12,393 + 71.1$
11. $666,773 \div 1,291$

12. −283 × 442

13. 19,532 − 4,860

14. 8.3 + 32,387

15. 34 + 484

Were you able to store the numbers and keep track of their respective operators? Don't worry if you struggled with this; it may take some time. Don't move on until you're confident.

Skill #3: Selecting a Method by Which to Solve the Problem

You need to select the appropriate solution method during all five phases of **Mental Calculation**. Once a problem is broken down into individual parts, stored, and assessed, you can determine whether any of the individual parts point you toward a specific method. If you have temporarily memorized 13 × 19, you should know to use the **Teeny-Teeny** or **FOIL** methods to solve it.

Once you've decided upon a method, you need to remember your selection as you go forward. You should be able to hold the selection in your head with relative ease—no mnemonic device needed.

Memorize the following problems. After memorizing each problem, choose a solution method and distract yourself with an unrelated task for a few minutes. Afterward, you should be able to recite a.) the problem (the individual numbers and operation involved) and b.) the chosen solution method. As before, don't worry about solving the problems at this time.

1. 34×111
2. $59 \div 8$
3. 9×610
4. $670 - 582$
5. $553 + 692$
6. $58 \div 11$
7. $1,200 \div 43$
8. 16×14
9. $780 - 170$
10. $17 + 901$

Multiple Operators

When you encounter problems that contain multiple operators, make sure the problem is as simple as possible before attempting to solve it. Let's illustrate this with examples:

$77 + 92 + 398 + 4$: This problem involves only addition. Store all the numbers in separate **Journeys** and store the operator in your natural working memory.

$88 - 44 - 21 - 6$: This problem involves only subtraction. You know by now that this is really only a two-part problem. In such a case, you want to break the problem down into multiple smaller problems. The first one will determine the **Subtraction Subtotal**. Store it and solve it. The second will subtract this **Subtraction Subtotal** from the **Master Value**. Store this as well and solve it.

$610 + 2,171 - 14$: Again, you want to solve one problem (the easiest being the addition problem), store the answer, create a new problem (the answer to the first problem minus 14), store this new problem, and solve it.

These are just the first three skills out of the seven needed to handle an unassisted, end-to-end **Mental Calculation** problem. Your adventure has just begun. Move on only when you're ready.

Quick Review

1. The first three skills required to successfully solve **Mental Calculation** problems are 1.) storing the problem, 2.) isolating the individual parts of the problem, and 3.) selecting a solution method.

Lesson 5: Mental Calculation—The Seven Skills (Part 2)

The last lesson covered the first three skills required to be able to solve end-to-end math problems using **Mental Calculation**. You're about to get intimately acquainted with the fourth.

Skill #4: Solving for the Individual Digits of the Answer

Obviously, the need to actually solve the problem applies to all five phases of **Mental Calculation**. Once you have everything you need (the problem's individual numbers, the operation, and your solution method) stored in the wonderful, low-voltage piece of meat that is your brain, you can begin to isolate and act on the parts of the problem to produce the individual digits of the answer. Let's try an easy example first.

The problem 12 × 13,465 is just a bit too long to store entirely in working memory, so the numbers can be stored within short **Journeys**. You should recognize that the **Trachtenberg System** method of multiplication by the number 12 is probably the simplest and most appropriate way to solve this problem.

This is the beauty of the **Trachtenberg System** methods for n × 5, 6, 7, 8, 9, 11, and 12: you can now forget about the number 12 in this problem and simply focus on the multiplier, 13,465. That's one less **Journey** to worry about.

Turning your attention to 13,465, begin solving the problem from the rightmost digit, as is customary with **Trachtenberg System** multiplication. Assuming you didn't have the problem in front of you, you would have already stored the entire number in your mind, and would now mentally traverse your **Journey** until its final **Step**. The **Mnemonic Clue** located there contains the rightmost digit of the number, which is 5.

The chosen method (**Trachtenberg System** multiplication by 12) dictates that you multiply this 5 by 2 and add its neighbor (which is 0, as 5 is the rightmost digit). This produces 10, which means 0 (the final digit of the answer) with a carried 1. For now, write down the 0. Storing answer digits in your mind is something you'll need for later phases, and we'll learn that later.

The next (second-to-last) problem digit must now be extracted and acted upon. Again, traverse the **Journey** until you come upon it. If using **Single-Digit Mnemonic Clues**, this will be found in the second-to-last **Step**.

Decipher this **Mnemonic Clue** into a number (6) and act upon it. This 6 must be multiplied by 2 and added to its neighbor (6 × 2 = 12, and 12 + 5 = 17). Add the carried 1 from the previous step (17 + 1 = 18). This leaves you with an 8 as the second-to-last answer digit and another carried 1. Write this 8 down to the left of the 0.

In many cases, you need to be aware of multiple digits at once. Here, not only do you have to act upon the 6, but you must also use the neighboring 5 to perform the calculation. By now you should be quick enough with number-to-mnemonic and mnemonic-to-number translations that both the mnemonic and number are interchangeable representations of one another. If so, performing these calculations by directly engaging the mnemonics themselves shouldn't be too difficult. With practice and patience comes comfort, but if mnemonic/number translation feels like it's using a lot of brainpower, you may want to work on your **Single-Digit Mnemonic Clues** a bit more.

Finish the example on your own. The final answer is 161,580.

Let's try a more difficult example: solving 91,484 × 311 using the **Two-Finger Method**. Given the number of digits involved, you'll need to use two **Journeys** from which you can extract the numbers as needed. First, memorize the problem.

Next, solve for the individual digits of the answer. To use the **Two-Finger Method**, add three 0s to the beginning and end of 91,484, making it 00091484000. This is necessary because there are three digits in the multiplier; if this confuses you, reread *Multiplying Large Numbers by Two-Digit Numbers*.

"But wait," you may say, "If I've already created my **Mnemonic Scenes** within my **Journey Steps**, how can I place three zeros at the beginning and end of the stored number?" There's no easy answer. Luckily, it's rare in non-competitive situations for problems of this magnitude to require precise answers. Simply use your working memory to remember the imagined zeroes before and after the number. There's no real need to add these digits to your existing **Journey**.

Go as slowly as you must, but solve for each individual answer digit using only your mind. Write the answer down digit-by-digit as you unveil it. This is hard. Be patient with yourself.

Once you finish, tackle another problem: $11,800 \times 24$.

If there's one lesson you should take away from this book, it's to be flexible and creative; why multiply by 11,800 when you can multiply by 118 and then toss the removed 00 on the right-hand side of the final answer? This turns a seven-digit problem into a five-digit one that you can manage in your head without using **Journeys**. You can then solve 118×24 through Trachtenberg's **Two-Finger Method** or by multiplying 118×12 using **Trachtenberg's** $n \times 12$ method and then multiplying the product by 2. Once you're done, simply re-attach the two zeroes to the final answer. Store the problem and solve it on your own, again writing the answer down as you go.

Even though you're storing numbers using **Mnemonic Clues**, the work itself requires the translated numbers to exist temporarily in your imagination. You'll need to visualize actual numbers in your mind. This is where powers of concentration come in handy; relax and imagine the numbers in high contrast colors, as bold, bright digits on a dark backdrop, blocking out the world's distractions.

Arguably, the most challenging problems are those that involve keeping track of several chunks of work simultaneously. Consider

Trachtenberg System division problems like 412 ÷ 17. These are difficult because **Trachtenberg System** division requires three layers of numbers: the problem (top row), the **Working Figures** (middle row), and the **Partial Dividends** (bottom row). Working on any part of the problem means you have to be aware of all three on at least a superficial level. With time and practice comes familiarity; there are no shortcuts.

Try to do a few problems. Briefly look at each of these problems one at a time, but turn away after a few moments of study. Look at a wall or close your eyes, then work out the answers in your head one digit at a time. Each time you discover an answer digit, open your eyes and write it down. Refrain from looking back at the problem. Continue until you've come up with the entire answer. Check each answer before moving on to the next problem. The answers are listed below.

1. 11 × 1,822
2. 9 + 87 + 134 + 18
3. 18 × 16 (remember your **Shortcuts**; in this case, the **Teeny-Teeny**)
4. 6 × 245
5. 231 × 4 (again, remember your **Shortcuts**)
6. 1,232 ÷ 88
7. 7 × 121
8. 288 − 55
9. 39 × 31 (yet again, remember your **Shortcuts**)
10. 1,298 − 981
11. 28 × 546

Answers:

1. 20,042

2. 248
3. 288
4. 1,470
5. 924
6. 14
7. 847
8. 233
9. 1,209
10. 317
11. 15,288

Quick Review

1. The fourth **Mental Calculation** skill is solving for the individual digits of the answer.

Lesson 6: Mental Calculation—The Seven Skills (Part 3)

> "MATHEMATICS, RIGHTLY VIEWED, POSSESSES NOT ONLY TRUTH, BUT SUPREME BEAUTY—A BEAUTY COLD AND AUSTERE, LIKE THAT OF SCULPTURE, WITHOUT APPEAL TO ANY PART OF OUR WEAKER NATURE, WITHOUT THE GORGEOUS TRAPPINGS OF PAINTING OR MUSIC, YET SUBLIMELY PURE, AND CAPABLE OF A STERN PERFECTION SUCH AS ONLY THE GREATEST ART CAN SHOW. THE TRUE SPIRIT OF DELIGHT, THE EXALTATION, THE SENSE OF BEING MORE THAN MAN, WHICH IS THE TOUCHSTONE OF THE HIGHEST EXCELLENCE, IS TO BE FOUND IN MATHEMATICS AS SURELY AS POETRY."
>
> —BERTRAND RUSSELL, MATHEMATICIAN

How's that for a quote?

You just learned the fourth skill required to solve math problems using **Mental Calculation**. Now let's wrap things up by discussing the final three.

Skill #5: Temporarily Storing the Individual Answer Digits

Congratulations on making it this far! All your hard work is paying off, and though you may not yet be able to effortlessly solve heavy-duty math problems in your mind, you're very close to understanding everything you'll need to do so.

Temporary mental storage of the individual answer digits, as opposed to writing them down, is required for **Mental Calculation** phases three through five. To lose the pencil and paper, you must record the individual digits of the answer one-by-one as you uncover them. You'll accomplish this using mnemonics and the **Journey Method**.

When preparing to face a math problem, you'll need to choose an additional **Answer Digits Journey**. You must choose one that's different from those used to store the problem's numbers. It is in this **Answer Digits Journey** that you will store the answer.

When choosing your **Journeys**, you were taught to make use of the space or setting directly around you. You can certainly continue to use your immediate surroundings to select an **Answer Digits Journey**. Some people instead select a specific **Answer Digits Journey** that they constantly overwrite, while others may elect to choose an arbitrary **Journey** from a predetermined collection. The choice is yours, though you should keep in mind that constantly reusing a static answer **Journey** can cause unnecessary confusion.

Regardless, whenever you find yourself needing to perform quick **Mental Calculation**, select and prepare a **Journey** in which to record the answer digits.

For problems that need to be solved from right to left (such asTrachtenberg System multiplication problems), your **Answer Digits Journey** needs to be populated backward. This means that the first digit you unveil (the one farthest to the right) should be stored within the **Journey's** final **Step**, so the digits can later be extracted going forward.

That can be confusing, so let's clarify with an example; if solving the problem 35 × 11 using the **Trachtenberg System** *n × 11* method, you would discover the answer digits from right to left. This would be "five...eight...three." Assuming the use of **Single-Digit Mnemonic Clues**, you would place the "five" **Clue** in the third **Step** of the **Journey**, "eight" in the second, and "three" in the first. Then, when it's time to collect the answer, you'd imagine yourself walking forward through your **Journey** to collect the answer digits (first **Step**, second **Step**, third **Step**): "three hundred...eighty...five." If you're unsure about how many digits the answer will be, start with the final **Step** of the **Journey** and work your way backward; if it turns out you don't need the first few **Steps**, so be it.

Skill #6: Combining the Individual Answer Digits Into a Final Answer, and, if Necessary, Conveying the Answer to Another Party

So far, you've learned how to store and solve math problems, and temporarily store the individual digits of the answers using only your mind. But what do you do with these digits once you have them all?

Combining the answer's individual digits into a final answer is necessary for phases three through five, during which you must compile, record, and convey answers without external aid.

First, walk through the **Answer Digits Journey** and extract the individual digits, forming a final, combined answer. Regardless of whether the answer was uncovered from right-to-left or from left-to-right, always traverse the **Journey** forward. As mentioned, in right-to-left scenarios, the answer digits should have been stored backward to account for this. Always account for direction while *storing* the answer digits, as opposed to when *extracting* them.

More often than not, the answer will eventually need to leave your head. This can mean conveying the answer to a medium (recording it, writing it down, typing it, etc.), but it can also mean verbally conveying the answer to another party.

Imagine you're stranded on a tropical island and must build a shelter to survive. Luckily, you're stranded with your friend, Guapo Biblioteca, who just so happens to be a world-renowned architect and engineer. Moreover, you were fortunate enough to have crashed while aboard a ship full of home building supplies; things could definitely be worse. After a few days of milling about and hoping to be rescued, reality has set in—it's time to begin building a home. Armed with the supplies and his experience, Guapo sets out to build not a mere shelter, but a tropical mansion.

Having been spoiled by the tools of his trade and the conveniences of home, Guapo is at a loss; he has no computer or calculator with which to perform his calculations. Having invested in a life-changing book by the name of *Master the Language of the Universe*, you assure him you can help.

While you clumsily attempt to harvest coconuts at the top of a tree, Guapo is below, busily drafting blueprints and treating you like a human calculator. He says, "The square footage of the foyer would be... what's 38.8 × 21.5?"

You think for a few moments and respond, "eight hundred thirty four point two." Amidst the loud tropical winds and howling monkeys, such a "plain English" answer is much less confusing than saying, "Okay, from right to left, it's 2, 4, 3, 8. And, uh...the decimal place is one place in from the right, between the 2 and the 4."

This may be an extreme, facetious example, but the point is that in the real world, your hard-earned ability to perform **Mental Calculation** is far less practical (and far less impressive) if you can't express the answer in a conversational, natural way.

Let's look at an example of a right-to-left problem: 21 × 188, solved via Trachtenberg's **Two-Finger Method** (the answer is 3,948). Reaching the end, you've placed 8, then 4, then 9, then 3 in the **Steps** of your **Journey**, travelling backward. By this time, you're aware that you used four **Steps**, and so the answer is four digits long. Thus, the format of the answer is X,XXX, or "X thousand, X hundred and X-X."

As you travel forward through your **Journey** to decode the answer of 3,948, you'll see the 3, and say "three thousand," then the 9, and say "nine hundred," then the 4, "forty-", and the final eight, "eight." You're able to deliver natural-sounding responses that are parsed into commonly understood linguistic patterns for numerical information (taking into account the language being used, of course).

Record each of the numbers listed below within the **Steps** of short **Journeys**, then retrieve them and announce the whole answer aloud in a conversational manner. Use different **Journeys** for each to limit confusion. Treat each number as the answer to a problem that's revealed left-to-right (such as **Trachtenberg System** division), so you'll record the digits moving forward through the **Journey**, and retrieve them the same way.

1. 12478
2. 9.73
3. 871
4. 82
5. 489032

With the second set of numbers listed below, you'll again record each within the **Steps** of short **Journeys**, then retrieve them and announce the whole answer aloud in a conversational manner. This time, treat each number as the answer to a problem that's revealed from right-to-left (such as **Trachtenberg System** multiplication). Record the digits in the order they're presented, moving backward through the **Journey**, beginning with the final **Step**. Then, retrieve the digits while moving forward through the **Journey**, beginning with the first occupied **Step**. For example, if faced with 34951, you'd record 3 in the final **Step** of the **Journey**, 4 in the penultimate, 9 in next, etc. You'd then imagine walking forward through the **Steps**, saying, "fifteen thousand…nine hundred…forty…three" aloud.

1. 8272
2. 291
3. 18374
4. 737182
5. 3297800

Once you're comfortable with this, try some full-blown problems. Look at the problems for a second or two, record them in your mind, solve them in your mind and announce the answer in a natural, conversational manner. The answers are below.

1. 77 × 192
2. 892 × 21
3. 762 − 181
4. 324 ÷ 4
5. 398 × 3,201

Answers:

1. "Fourteen thousand, seven hundred and eighty four."
2. "Eighteen thousand, seven hundred and thirty two."
3. "Five hundred eighty one."
4. "Eighty one."
5. "One million, two hundred seventy three thousand, nine hundred and ninety eight."

Stop here and work on this skill for as long as you need.

Skill #7: Storing the Answer

You may not always find yourself interacting with another person or writing your answers down when solving math problems; you may instead be performing calculations to glean information from them for later access. In these cases, you will want to store the final, combined answers in your head without recording them to external media or needing to announce them aloud.

If you must store a problem's final answer for any significant period of time, it's best to convert the answer into a series of **Single-Digit Mnemonic Clues** and store them in a **Journey**. Raw numbers are easily forgotten, while interesting and creative **Mnemonic Scenes** are vastly more memorable. If the individual digits have been stored in a forward direction (with the leftmost answer digit in the first **Journey Step**), you can use this **Journey** to store the information indefinitely; however, if the individual digits were initially stored backward, place them in a new **Journey** that hasn't yet been used for this problem. In short, make sure the **Mnemonic Clues** are stored in a forward direction within your chosen **Journey** so you can easily retrieve them later.

If you do need to use a new **Journey**, try to choose one that makes sense given the subject matter (something known as a **Contextual Journey**). If you've determined that the square footage of your basement is five hundred and forty three square feet, you may want to store this number in a **Journey** set in said basement. Relating your **Journey** to the topic drastically increases your chances of recalling exactly which **Journey** holds this information, minimizing confusion.

Putting It All Together

It's important to remember that real-life problems don't often require high levels of precision, so you can often skip the hardest parts of **Mental Calculation**. Let's walk through a realistic and practical end-to-end problem. You're considering changing your work 401(k) or benefits plan so that you'll save $6.31/day for the remaining 185 days of the year. About how much money will you save?

The precise correct answer is $1,167.35. How close can you get without doing too much work?

You could drastically simplify the problem to 63 × 2 and multiply the product by 10. The result is $1,260, which is off by about $100. This would be sufficient for most purposes; it paints a clear picture of the type of money you'd be saving. You can do this in your head very easily.

If you wanted a bit more precision, you could multiply 63 × 18 using **FOIL**. This would give you $1,134, which is closer.

Given it's size, you can most likely memorize the problem with your working memory instead of using a **Journey**. If doing it this way, you can imagine the problem written out, apply **FOIL**, and keep a running tally of the answer total in your mind. "six hundred...one thousand eighty...one thousand, one hundred ten...one thousand, one hundred thirty-four."

In most practical situations, it would be reasonable to sacrifice a little bit of precision to avoid using the **Trachtenberg System**. The real challenge arises when you require true precision. Let's work through another problem together, this time with the utmost precision: 48 × 2,935.

You first need to memorize the problem and chosen solution method, so look around your surroundings and select two **Journeys**. The first will house the **Single-Digit Mnemonic Clues** for 2, 9, 3, and 5, and the second will house the **Single-Digit Mnemonic Clues** for 4 and 8. Commit the operator (multiplication) and solution method (**Trachtenberg Two-Finger Method**) to your working memory so you no longer have to refer to the problem. Select a third **Journey** in which to store your answer digits, clear your mind, and begin.

Imagine performing the required calculations between these two numbers, keeping in mind the two imaginary 0s to the left and right of the larger number. As you reveal each answer digit, begin filling your **Answer Digits Journey**, starting with the final **Step** and working backward. Once you're finished, figure out how many answer digits there are (based on the number of **Steps** used), and then imagine walking forward through your **Journey**, beginning with the first occupied **Step**. Collect the answer digits: "One hundred… forty… thousand… eight hundred …eighty." If you'd like to store this number in your head for later use, choose yet another **Journey** and use it to store the digits in a forward direction, starting with the first **Step**.

This seems like quite a few steps, but this is a brand-new skill. Your mind is an incredible machine, and with a lot of practice, it's surprising how quickly you will be able to solve these types of problems. **Mental Calculation** will always be a challenge, but these methods create a clear blueprint for you to follow.

Quick Review

1. The final three skills required for successfully solving a **Mental Calculation** problem are 1.) temporarily storing the individual answer digits, 2.) combining the individual answer digits into a final answer, and if necessary, conveying the answer to another party, and 3.) storing the answer.

Conclusion

The most important thing to remember about **Mental Calculation** is that there are no magic shortcuts. You have all the tools you need, from the skills needed to perform calculations to the methods used to store them entirely in your mind. However, you will still need to work diligently to fully develop this ability; if it's worth it to you, put in the hours.

Let's review the **Five Phases of Mental Calculation** and their respective skills all together, now that you understand everything involved. Like before, the aspect that is new to each phase (compared to the previous) is presented in *italic*.

1. Reference the problem on paper (or screen, etc.) as needed, perform your work on paper, record the answer on paper, and announce the answer. <u>Skills</u>: 1.) Selecting a solution method, and 2.) calculating the individual digits of the answer.
2. Reference the problem on paper as needed, *do the work in your head*, record the answer on paper, and announce the answer. <u>Skills</u>: 1.) Isolating the individual parts of the problem, 2.) selecting a solution method, 3.) calculating the individual digits of the answer, and 4.) temporarily storing the individual answer digits.
3. Reference the problem on paper as needed, do the work in your head, *record the answer in your head*, and announce the answer. <u>Skills</u>: 1.) Isolating the individual parts of the problem, 2.) selecting a solution method, 3.) calculating the individual digits of the answer, 4.) temporarily storing the individual answer digits, and 5.) combining the individual answer digits into a final answer and conveying the answer to another party.

4. *Briefly view the problem briefly on paper (without continuing to reference it)*, do the work in your head, record the answer in your head, and announce the answer. <u>Skills</u>: 1.) Storing the problem, 2.) Isolating the individual parts of the problem, 3.) selecting a solution method, 4.) calculating the individual digits of the answer, 5.) temporarily storing the individual answer digits, and 6.) combining the individual answer digits into a final answer and conveying the answer to another party.

5. *Listen to the problem (verbally)*, do the work in your head, record the answer in your head, and announce the answer. Alternately, store the answer in your head for later retrieval. <u>Skills</u>: 1.) Storing the problem, 2.) Isolating the individual parts of the problem, 3.) selecting a solution method, 4.) calculating the individual digits of the answer, 5.) temporarily storing the individual answer digits, 6.) combining the individual answer digits into a final answer and conveying said answer to another party, and 7.) storing the answer.

Review and Development: Section 6

You now know how to solve math problems using **Mental Calculation**, and should be equipped to solve most practical day-to-day problems; however, this doesn't mean you should already be quick. Like all impressive or worthwhile endeavors, expertise is the result of time, patience, and practice. Here, in development, you'll truly begin to cultivate this skill.

Review

1. Numbers are a good example of mundane information, which your brain tends to discard. In order to retain numbers, you must associate each digit with an imaginative **Clue**, which your brain can usually retain much more easily.

2. A **Mnemonic Vocabulary** is a personalized set of **Mnemonic Clues** that represent specific pieces of mundane information.

3. Creating **Single-Digit Mnemonic Clues** (standardized images or objects that represent each number from 0 through 9) allows you to easily imagine and recall individual digits. Some of the most popular **Single-Digit Mnemonic Clues** are those that either physically resemble ("2" and "a cobra") or sound like ("three" and "tree") the digit.

4. A **Journey** is an imagined path through a familiar setting in which you store information. You can do so by integrating **Mnemonic Clues** (such as **Single-Digit Mnemonic Clues**) within the individual **Steps** that comprise the **Journey**. You can then mentally retrace the **Steps** of the **Journey**, observing the

Mnemonic Clues and translating them back into information (such as numbers).

5. <u>Focus/Concentration</u>: **Focus/Concentration** refers to the ability to stay completely engaged with a problem (and its solution) amidst distractions. In order to develop this skill, you should progress through three levels of increasing difficulty: First, learn to concentrate on *nothing*; then, learn to concentrate on *something*; finally, learn to concentrate on *several things at once*.

6. <u>Mental Calculation</u>: **Mental Calculation** refers to the act of solving arithmetic problems in your head, without the aid of calculators, your fingers, or any other tools, including analog tools like a pencil and paper. **Mental Calculation** is comprised of three components: ability (calculation methods), memory, and focus/concentration.

7. <u>The 5-7 Approach</u>: The **5-7 Approach** to **Mental Calculation** refers to five phases and seven skills.

8. <u>The Five Phases</u>: There are five phases of **Mental Calculation**, each a bit more challenging and impressive than the last. Reaching the fifth and final phase of true, unassisted **Mental Calculation** will require all seven distinct **Mental Calculation** skills outlined below.

9. <u>The Seven Skills</u>: The seven skills required for successfully solving a **Mental Calculation** problem are 1.) storing the problem, 2.) isolating the individual parts of the problem, 3.) selecting a solution method, 4.) solving for the individual digits of the answer, 5.) temporarily storing the individual answer digits, 6.) combining the individual answer digits into a final answer, and if necessary, conveying the answer to another party, and 7.) storing the answer.

Development

Spend seven days working on **Mental Calculation** as covered in *Section 6.*

Day 1

Start by ensuring you've memorized the five phases and seven skills involved in **Mental Calculation**. By the end of the first day, you should be able to recite them all by memory. You should also understand the meanings behind each sufficiently enough that you could explain them to someone if prompted.

Days 2 – 3

Begin with the phase of **Mental Calculation** you're currently comfortable with and push your boundaries: attempt to solve problems within the parameters of the next phase. Practice deliberately, patiently, and without worrying about how long it takes to solve each problem. Make up arbitrary problems on your own (addition, subtraction, multiplication and division).

Days 4 – 5

By the fourth day, you should be operating at either the fourth or final phase, and able to solve math problems in your mind after hearing the problem said aloud only once. *It's fine if you're still slow*; speed is less important than technique or precision. If your situation permits, have someone verbally provide you with problems and check your answers after you solve each one.

Days 6 – 7

Over the last two days, time yourself and proactively try to improve your solution speeds. Be sure to compare speeds between problems of similar types and similar lengths, as some tend to take longer than others. Set several measurable goals for the next month, next three months, and beyond. For example, you could make it a goal to be able to solve 'five-digit multiplied by two-digit' problems using Trachtenberg's **Two-Finger Method** within one minute by a certain date. Once you've settled upon a goal, construct a personal practice plan that will get you there.

In the next section, *Advanced Calculation*, you will venture into a realm of mathematical prowess that transcends pretty much all practical applications.

Section 7: Advanced Calculation

Math is the language of the universe, and most of us are less than fluent in it. Numbers are words in an endless conversation between humanity and the cosmos, allowing us to connect with it in some small way.

You are now considerably more fluent in this language.

By now, you are capable of **Mental Calculation** feats that far exceed the requirements of most common problems. If you'd like to transcend the practical, cross the line from impressing people to astonishing them, and strive to become a world-class **Mental Calculator**, every tool you'll need (aside from discipline) can be found in this section.

You'll begin with addition, and then move on to advanced multiplication and division. The following is not for the weak of spirit. Begin if you dare.

Lesson 1: Advanced Addition and Subtraction

In committing to tackle the most challenging and awe-inspiring lessons of this book, you have indeed entered the realm of the most excellent. You will expand upon your existing skills and perform jaw-dropping feats of advanced **Mental Calculation**. We will begin, as usual, with addition and subtraction.

Combining Left-To-Right Addition/Subtraction with Familiar Numbers

You know how to add numbers from left to right. You also know how to use **Familiar Numbers** to make complex problems more manageable. It's time to combine these two ideas; like peanut butter and jelly, they are good on their own, but almost magical when paired. You can take advantage of **Pre-Scanning** and round down in such a way that you can often reduce **Overflow Risk** and therefore make problems easier.

While developing this skill, write your problems down. Even with your newfound **Mental Calculation** abilities, the amount of concentration required to remember both numbers can distract you from the task at hand, and you want to focus on the method. Keep a pencil and paper handy throughout this process.

Before bringing **Familiar Numbers** into the picture, let's approach a problem in the left-to-right manner you already learned.

Let's use 53,181 and 22,601. **Pre-Scanning** the first number (53,181) reveals that only the 8 is large enough to pose an **Overflow Risk**. Assessing the second number (22,601), concentrate on whether or not the 8's corresponding digit will result in 9 or higher. Since it's a 0, it doesn't, and no other digit in the second number poses an obvious **Overflow Risk**, so you're clear to add from left to right with no interruptions.

If that made no sense to you, go back and review the basics of addition as described in *Section 1: Fundamental Mathematical Concepts*.

Now, combining this with the concept of **Familiar Numbers**, round 53,181 to 53,180 and 22,601 to 22,600 before adding left-to-right. This results in an easier addition problem (53,180 + 22,600). Once you've added the two numbers together, simply add the resulting **Offset** of +2. Note that both numbers were rounded down, as each contained ones place digits with values less than 5.

Let's look at another example, this time with an obvious **Overflow**: 21,364 + 12,182. Before adding left-to-right, round 21,364 to 21,360, and 12,182 to 12,180. Do a quick left-to-right **Pre-Scan** of the two rounded numbers to look for **Overflow Risks**; if you find none, quickly add from left to right. However, the first number (21,360) triggers an alert at the tens place, as the 6 (being greater than 5) presents a potential **Overflow Risk**. A scan of the second number shows that the 6's counterpart (the other number's tens place) is an 8, and thus these two digits will **Overflow**. Seeing this lets you know that you can't simply add; at least one **Overflow** must be accounted for. Begin adding the digits starting from the far left. 2 + 1 = 3; then, 1 + 2 = 3; then, 3 + 1 = 5 (as there's a definite **Overflow** in the neighboring pair), then, 6 + 8 = 4, and finally, 0 + 0 = 0. Add your **Offset** of +6 to arrive at the final answer, 33,546.

Finally, let's try an example with less obvious **Overflow Risk**: 22,649 + 35,350. Write these numbers down and work through the problem on your own. How will you approach it? Feel free to speak aloud as though explaining the process to someone (even better, give a friend or family member an impromptu lesson; they'll likely be a bit confused and bored, but they'll live). When you're finished, let's break down the process.

How did you round the numbers? Even though 22,649 is closer to 22,650 than 22,640, you should have rounded down. If you rounded up, the 649 at the end of 22,649 would have become 650, which introduces **Overflow** unnecessarily; the two added 5s (from 22,6<u>5</u>0 and 35,3<u>5</u>0) would exceed 9, while 4 plus 5 (from 22,6<u>4</u>0 and 35,3<u>5</u>0) wouldn't. Rounding down, 649 turns into 640, and the 350—already a **Familiar Number**—is left alone. While **Pre-Scanning**, you should have noticed that the 6 and 3 (the hundreds place digits) add up to 9, and therefore pose an **Overflow Risk**; so, you must look to the neighboring pair of digits (the tens place) to determine whether an **Overflow** will take place. Since the rounded digits of the neighbor pair are 4 and 5, which equal 9, you're safe. The rest of the problem is somewhat straightforward; you must simply add (from left to right) and account for the **Offset**.

Now let's do this problem again, but with a slight difference that will make it a bit harder; let's turn 22,649 + 35,350 into 22,649 + 35,360. Again, write these numbers down and walk through the problem.

Even though 22,649 is closer to 22,650 than it is to 22,640, you should round down, as you're trying to minimize **Overflow**. Unfortunately, due to the nature of the problem, **Overflow** will occur no matter what: 4 plus 6 (from 22,6<u>4</u>0 and 35,3<u>6</u>0) exceeds 9. Had you rounded up, the 649 at the end of 22,649 would have become 650; you'd be introducing even worse **Overflow**, as 5 plus 6 (from 22,6<u>5</u>0 and 35,3<u>6</u>0) would reach 11. Rounding down, you turned 649 into 640, and the 360—already a **Familiar Number**—gets left alone.

While **Pre-Scanning**, you should have recognized that the 6 and 3 (the hundreds place digits) add up to 9, and therefore present an **Overflow Risk**. So, you must look to the neighboring pair (the tens place digits) to determine whether or not an **Overflow** will indeed occur: 4 + 6 = 10, so there *will* be an **Overflow**, and it will be passed to the hundreds place before finally being passed to the thousands place. Quite a mess.

The rest is handled as usual—simply add (from left to right) and account for the **Offset**.

This messy, ugly problem illustrates a simple rule; when dealing with larger problems, it's often safer to round down than up. **Overflows** are more likely with larger numbers than they are with smaller ones; the more digits, the more opportunities for **Overflow**. Shrinking a digit—

even the ones place—reduces the chance of an **Overflow**, so *round down whenever possible.*

This problem also demonstrates that you can combine what you already know into a three-pronged approach toward any addition problem, regardless of complexity:

1. Round the numbers to **Familiar Numbers** and try to tactically avoid **Overflows** whenever possible.
2. Add the rounded numbers (left-to-right).
3. Adjust for the **Offset**.

Subtraction Using the Same Concepts

Subtraction is performed the same way; round the numbers, subtract from left to right, and account for the **Offset**. Piece of cake. Keep in mind you must calculate the **Offset** in subtraction problems using **Dual Offset Subtraction**. The direction of rounding is therefore of no consequence, as long as both numbers are rounded in the same direction.

Let's do an example: 11,189 – 8,721.

Both 89 and 21 (from 11,1**89** – 8,7**21**) are the same distance from the nearest tens place, so you can easily round both numbers down. 11,1**89** becomes 11,1**80** (a difference of 9) and 8,7**21** becomes 8,7**20** (a difference of 1). Scanning the rounded problem (11,180 – 8,720) from left to right, you first see that 11 – 8 = 3; but then you see that the next step (1 – 7) requires you to borrow, so the 3 (from 11 – 8) is actually reduced to a 2. You can solve 1 – 7 = 4, since you borrowed from the left to make this $11 - 7 = 4$. Next, finish the problem; $8 - 2 = 6$, and finally, $0 - 0 = 0$. The pre-**Offset** answer is 2,360. There's an **Offset** of -8. Subtracting this from the pre-**Offset** answer provides the final answer of 2,468.

These basic principles should apply to any subtraction problem involving two numbers, regardless of the level of complexity.

Do the following problems once with pencil and paper, and then again using only your mind (employing the **Mental Calculation** methods you learned). When using **Mental Calculation**, resist the urge to panic and guess; take your time and solve with confidence and composure.

1. 123,873 + 67,219
2. 2,387 + 23,987,298
3. 6,574,839 + 10,293,847
4. 9,872 + 239,821
5. 34,897 + 780,234
6. 4,532 – 120
7. 5,198 – 3,290
8. 2,341 – 349
9. 789,234 – 2,343
10. 9,872,345 – 87,923

Answers:

1. 191,092
2. 23,989,685
3. 16,868,686
4. 249,693

5. 815,131
6. 4,412
7. 1,908
8. 1,992
9. 786,891
10. 9,784,422

Quick Review

1. The principles of left-to-right addition/subtraction and
   Familiar Numbers can be combined into a three-step process
   for quickly solving two-number addition or subtraction
   problems. First the numbers are rounded to **Familiar Numbers**
   in such a way that you avoid **Overflows** whenever possible.
   Second, the problem is solved from left to right. Finally,
   account for any **Offset**.

Lesson 2: Multiplying Large Numbers by Other Large Numbers

> "MATHEMATICS IS ONE OF THE DEEPEST AND MOST POWERFUL
> EXPRESSIONS OF PURE HUMAN REASON, AND, AT THE SAME TIME,
> THE MOST FUNDAMENTAL RESOURCE FOR DESCRIPTION AND
> ANALYSIS OF THE EXPERIENTIAL WORLD."
>
> —HYMAN BASS, MATHEMATICIAN

You learned earlier how to multiply any number, regardless of size, by a two-digit number with relative ease and grace using the **Trachtenberg System**. Assuming you became appropriately comfortable with the method, you should now be well equipped to handle such problems. The multiplier, however, won't always be just two digits.

To multiply large numbers by other large numbers, you must simply extend what you've already learned. You'll begin by multiplying large multiplicands by three-digit multipliers, and then learn how to scale the method as needed.

Problems Involving Three-Digit Multipliers

When solving a problem involving a three-digit multiplier, stuff three zeroes before and after the multiplicand. Example: 2,134 × 461.

> 2,134 × 461

becomes...

> **000**2,134**000** × 461

It's probably obvious that when this method is scaled to and beyond four-digit multipliers, you'll add one zero to each side for each digit in the multiplier. For a ten-digit multiplier, this means you need to place ten zeroes on each side of the multiplicand. This is a daunting test of your **Mental Calculation** abilities, but luckily, numbers that large don't come up often in daily life—and if they do, they rarely require precision in cases where calculators aren't available.

When faced with a three-digit multiplier, the **Inside Pair** is calculated through the connection of the leftmost (hundreds place) digit of the multiplier (in your case, the 4 from **4**61), and the two rightmost digits of the multiplicand (the 00 from 0002,134**00**). First calculate 4 × 0 (the 0 from **0**0) and 4 × 0 (the 0 from 0**0**), and add the results to arrive at your first **Pair-Product**: 0 + 0 = 0. Hold onto this 0 going forward into the next step.

If you're feeling unsure about some of this terminology, go back and review what you learned about multiplication in *Section 5*.

> **Inside Pair:** 0 (0002,134**00** × **4**61)

You're familiar with the **Inside** and **Outside Pairs,** but with a three-digit number, there's a new term to learn: the **Middle Pair.** Continue to work your way through the multiplicand from right to left. You already handled the 4 (4**6**1) with the **Inside Pair,** so move on to the 6 (4**6**1). Your focus within the multiplicand shifts to the left one place (from 002,134**00** to 002,134**000**). Using these numbers, you can determine the **Pair-Product** for the **Middle Pair.**

Take the middle digit of the multiplier (the 6 from 4**6**1) and the 00 from 0002,134**000**. Math time: 6 × 0 (the 0 from **0**0) = 0, and 6 × 0 (the 0 from 0**0**). You now have the second **Pair-Product:** 0 + 0 = 0. Hold this in your mind along with the first 0 you've been storing (the **Inside Pair's Pair-Product**).

> **Middle Pair:** 0 (0002,134**000** × 4**6**1)

Finally, the **Outside Pair's Pair-Product** is calculated through the connection of the rightmost (ones place) digit of the multiplier (the 1 of 46**1**) and the next two digits in the multiplicand (after another one-digit shift to the left). Use the same familiar method to multiply 1 (from 46**1**) against the 40 (from 0002,13**40**0).

1 × 4 = 4 and 1 × 0 = 0. 4 + 0 = 4, which is the **Pair-Product** for the **Outside Pair:**

> **Outside Pair:** 4 (0002,13**40**0 × 46**1**)

To make sure this is clear, let's visualize all the pairings side-by-side.

> **Inside Pair:** 0 (0002,134**00** × 4**6**1)
> **Middle Pair:** 0 (0002,134**00** × 4**6**1)

353

Can you see how the numbers are connected with an imaginary arch that touches first the inside, then the middle, and finally the outermost parts of the numbers? Each pair involves numbers that are progressively farther from each other. You now have three **Pair-Products** stored in your head (0, 0, and 4); add them to arrive at the ones place answer digit.

4 + 0 + 0 = 4. This is the first answer digit.
The answer so far: 2,134 × 461 = xxxxx4

While collecting the **Pair-Products** you need to add together to find the answer digit, it may seem intuitive to hold onto each and add them all together at the end. It's actually considerably more efficient to add them as you go.

In the example, you first arrive at the **Inside Pair's Pair-Product**, 0. When you arrive at the second 0 (the **Pair-Product** of the **Middle Pair**), you can add 0 to the first 0 to create a running total of 0. Finally, add the **Outside Pair's Pair-Product** (4) to the running total (0), giving you a final answer digit of 4.

Adopting this practice means you only have to keep track of one number (the running total) at any given time.

Much as with two-digit multipliers, you need to continue moving from right to left until you run out of usable numbers, which is when your innermost multiplier digit is pitted against only zeroes. Let's finish this problem together. The next set of three pairs is:

Inside Pair: 0 (0002,134**00**0 × **4**61)

Middle Pair: 4 (0002,134**0**00 × 4**6**1)

Outside Pair: 3 (0002,1**3**4000 × 46**1**)

0 + 4 + 3 = 7. This is the second answer digit.

The answer so far: 2,134 × 461 = xxxx74

then...

Inside Pair: 6 (0002,134**0**00 × **4**61)

Middle Pair: 10 (0002,1**3**4000 × 4**6**1)

Outside Pair: 1 (0002,**1**34000 × 46**1**)

6 + 10 + 1 = 17 (7, **carry the 1**). This is the third answer digit.

The answer so far: 2,134 × 461 = xxx774

then...

Inside Pair: 3 (0002,1**3**4000 × **4**61)

Middle Pair: 7 (0002,**1**34000 × 4**6**1)

Outside Pair: 2 (000**2**,134000 × 46**1**)

3 + 7 + 2 + (the carried) 1 = 13 (3, **carry the 1**). This is the fourth answer digit.

The answer so far: 2,134 × 461 = xx3774

then...

Inside Pair: 5 (0002,**1**34000 × **4**61)

Middle Pair: 2 (000**2**,134000 × 4**6**1)

Outside Pair: 0 (00**0**2,134000 × 46**1**)

5 + 2 + 0 + (the carried) 1 = 8. This is the fifth answer digit.

The answer so far: 2,134 × 461 = x83774

then...

then...

If you were to keep going, none of your calculations would touch any of the multiplicand's actual digits (only 0s), so you're finished. The answer is 983,774.

Tip: *Whenever there are zeroes at the end of a number (such as 77,000 or 2,100), you can temporarily ignore them during the calculation steps and simply add them back onto the answer at the very end. This goes for both multiplier and multiplicand. This quick (and somewhat obvious) shortcut should be taken into consideration when solving large multiplication problems. For example, look at 19,524,000 × 1980. You can multiply 19,524 × 198 (which gives you 3,865,752), collect the zeroes (four of them in total), and stick them at the end of the answer (38,657,520,000).*

Problems Involving Multipliers More Than Three Digits in Length

There's no secret to multiplying larger numbers. The method you just learned is scalable:

1. For each digit of the multiplier, you must add a 0 on both ends of the multiplicand.
2. The more pairs there are between the **Inside** and **Outside Pair**, the more **Pair-Products** you must collect in order to arrive at each answer digit. This means multiple **Middle Pairs**.

Let's begin a problem together:

44,321 × 55,010

We won't walk all the way through it, as that would take up an exorbitant amount of book real estate.

A quick note: If you often find yourself needing to solve problems like 44,321 × 55,010 on your own, by all means, use a calculator. This is in no way a practical skill (but it certainly is impressive).

This won't be pretty, so take a deep breath.

44,321 × 55,010
...becomes...
000044,321**00000** × 55,010

You could peel the rightmost 0 off of 55,010 as discussed, but for the sake of working through an example, let's leave it.

0000044,32100000 × 55,010
0000044,32100000 × 55,010
0000044,32100000 × 55,010
0000044,32100000 × 55,010
0000044,32100000 × 55,010

$0 + 0 + 0 + 0 + 0 = 0$
The answer so far: 44,321 × 55,010 = x,xxx,xxx,xx0

then…

0000044,32100000 × 55,010
0000044,32100000 × 55,010
0000044,32100000 × 55,010
0000044,32100000 × 55,010
0000044,32100000 × 55,010

$0 + 0 + 0 + 1 + 0 = 1$
Answer so far: 44,321 × 55,010 = x,xxx,xxx,x10

then…

0000044,32100000 × 55,010
0000044,32100000 × 55,010
0000044,32100000 × 55,010
0000044,32100000 × 55,010
0000044,32100000 × 55,010

$0 + 0 + 0 + 2 + 0 = 2$
The answer so far: 44,321 × 55,010 = x,xxx,xxx,210

then…

$$0000044,3210\underline{00}000 \times 5\underline{5},010$$
$$0000044,321\underline{0}0000 \times 5\underline{5},010$$
$$0000044,3\underline{2}100000 \times 55,\underline{0}10$$
$$0000044,\underline{32}100000 \times 55,0\underline{1}0$$
$$00000\underline{44},32100000 \times 55,01\underline{0}$$

$0 + 5 + 0 + 3 + 0 = 8$
The answer so far: $44,321 \times 55,010 = $ x,xxx,xx8,210

…etc.

We'll stop here for practicality's sake. If this is making sense to you, you'll have no problem figuring out the rest. The final answer is 2,438,098,210.

Do the following problems on your own. You can start with a pencil and paper, but after you fully understand and feel confident with the method, move all the work into your head. Don't write out the intermediate steps; only record the problem itself and the answer digits. Each problem may take a while to solve; that's fine. Take your time and focus.

1. $1,944 \times 760$
2. 450×128
3. $987,234 \times 3,241$
4. $7,861 \times 23,678$

Answers:

1. $1,477,440$

2. 57,600
3. 3,199,625,394
4. 186,132,758

Before moving on, take a few days to develop this skill and increase your comfort level. Practice as often as you can. Come up with arbitrary numbers to work with, or find numbers in the world around you, such as UPC numbers, the numeric portions of license plates, and phone numbers. Always begin by using a pencil and paper, and slowly transition all aspects to your mind over time, progressing through the **5 Phases of Mental Calculation**. This will require intense concentration and mental isolation, which come only with time.

When all is said and done, this method is pretty straightforward and simply involves adapting and scaling **Trachtenberg System** multiplication methods you already know well.

Quick Review

1. The **Trachtenberg System** provides a single method by which to multiply two large numbers together. This method scales in order to accommodate problems involving numbers with any number of digits.

Lesson 3: Trachtenberg Division by Three+ Digit Divisors

If you think solving lengthy division problems is fun, there may be something wrong with you.

Whenever precision isn't necessary, reduce large division problems to simpler terms by uniformly knocking a digit or two off the right sides of each number involved; however, if you'd like to learn how to divide large numbers with precision, you're in the right place. Earlier, you learned how to divide large numbers by two-digit numbers using the **Trachtenberg System**. You'll now learn how to attack problems involving two numbers that are each at least three digits long.

You've come this far; why stop now?

Three-Digit Divisors

Dividing by a three-digit number is done almost exactly the same way as dividing by a two-digit number, which you'll see as we go through an example problem: 91,560 ÷ 654. First, you'll learn about the subtle distinctions involved.

1. **The UT Calculation**: Whereas with two-digit divisors you used a **U Calculation**, you'll now use a **UT Calculation**. In fact, the **U Calculations** used with two-digit problems were just **UT Calculations** without a third divisor digit (the **T**, or tens part).

2. **Remainders**: With two-digit divisors, the remainder was the final **Partial Dividend**. Moving beyond two-digit divisors requires a different approach. First, determine where your answer (the quotient) ends and your remainder begins. To do this, subtract 1 from the number of digits in the divisor (in the current example, 654 has 3 digits, so subtracting 1 gives you 2), and place a little mark this many digits from the right side of your dividend. In this case, the dividend is 91,560, and so you would place this mark between the 5 and the 6. Any answer digits produced from calculations performed on the numbers to the left side of this mark belong to the quotient, and the remainder is handled through a new process you'll soon learn that involves only activity occurring to the right of this mark.

Let's begin the problem. To get the first **Partial Dividend**, bring the 9 down from the dividend.

91,560 ÷ 654 =	Problem
	Working Figures
<u>9</u>	Partial Dividends

6 (from the divisor) goes into this 9 only one time (this is the first digit of the answer).

> 91,560 ÷ 654 = <u>1</u>XX

Then perform the **NT Calculation**, like before:

> 1 (your answer digit) × 6 (from <u>6</u>54) = <u>06</u> (this is your **N**, so keep the whole product)
>
> 1 (your answer digit) × 5 (from 6<u>5</u>4) = <u>0</u>5 (this is your **T**, so only keep the tens place digit of the product).
>
> 06 (**N**) + 0 (**T**) = 6.

9 (**Partial Dividend**) – 6 (**NT Calculation** outcome) = 3.

91,560 ÷ 654 = 1	Problem
<u>3</u>	Working Figures
9 (-6)	Partial Dividends

Bring the next untouched digit (1) down from the dividend (9,156) to fulfill the second half of the first **Working Figure**, which then becomes 31.

91,560 ÷ 654 = 1	Problem
31	Working Figures
	Partial Dividends

So far, this has gone just as it would have with a two-digit divisor; you haven't gotten to the rightmost digit of the divisor (65**4**) yet.

Here's where things change. With a 2-digit divisor, you would have performed a **U Calculation**, and subtracted the outcome of this **U Calculation** from the **Working Figure** (31). Since this divisor has three digits, you must instead perform something called the **UT Calculation** (6**54** × **1**):

5 (from 6**5**4) × 1 (latest digit of the answer) = 0**5**. Keep the 5 (**U**).

4 (from 65**4**) × 1 (latest digit of the answer) = **0**4. Keep the 0 (**T**).

Note that this calculation begins with the same divisor digit as a **U Calculation**; if you were working with 54 instead of 654, the **U Calculation** would have simply involved the 4. The calculation now extends to use the middle digit of the divisor, meaning you have enough digits to calculate the **T**. As such, the **U Calculations** you performed when working with two-digit divisors were actually **UT Calculations** without a third divisor digit with which to perform the **T** part. This is called the **UT Calculation** because it involves the u̲nits

364

(ones) place of the first calculation and the tens place of the second. Together, 5 (**U**) and 0 (**T**) = 5. This is the number you will subtract from the latest **Working Figure** (31) to determine your next **Partial Dividend** (26).

91,560 ÷ 654 = 1	Problem
31 (-5)	Working Figures
26	Partial Dividends

Continuing on, 6 (from **6**54) goes into 26 about 4 times, so this will be the next answer digit.

91,560 ÷ 654 = 14X

Now repeat the cycle. Take this newfound answer digit (4), and use it to perform an **NT Calculation** against 654.

4 (latest digit of the answer) × 6 (from **6**54) = **24**. Keep the entire answer (**N**).

4 (latest digit of the answer) × 5 (from 6**5**4) = **20**. Keep the 2 (**T**).

24 (**N**) + 2 (**T**) = 26.

Subtract 26 (from the **NT Calculation**) from 26 (the latest **Partial Dividend**) to arrive at 0...

91,560 ÷ 654 = 14	Problem
<u>0</u>	Working Figures
26 (-26)	Partial Dividends

...which is paired with 5 from the above row to arrive at your next **Working Figure**, 05.

91,560 ÷ 654 = 14	Problem
0<u>5</u>	Working Figures
	Partial Dividends

From here, determine the next **UT** as you did above. Multiply the new answer digit (4) by the 5 and 4 of 654.

> 4 (latest digit of the answer) × 5 (from 6<u>5</u>4) = 2<u>0</u>. Keep the 0 (U).
>
> 4 (latest digit of the answer) × 4 (from 65<u>4</u>) = <u>1</u>6. Keep the 1 (T).
>
> 0 (U) + 1 (T) = 1.

Once you've uncovered two answer digits, something changes; on top of the **UT Calculation**, you must now also perform a **U Calculation** (U as in <u>u</u>nits), using the second-to-last answer digit (the 1 from <u>1</u>4) and rightmost digit of the divisor (the 4 from 65<u>4</u>). You'll understand why in a moment.

> 4 (from 65<u>4</u>) × 1 (from <u>1</u>4) = 0<u>4</u>. Keep the 4 (U).

Then add the **UT** and the **U** products together:

1 (**UT**) + 4 (**U**) = 5.

Subtract this number from the latest **Working Figure** (5) to obtain your next **Partial Dividend** (0).

91,560 ÷ 654 = 14	Problem
05 (-5)	Working Figures
<u>0</u>	Partial Dividends

You may be wondering why you had to perform a **U Calculation** here. It's actually a **UT Calculation**, but there aren't enough answer digits to perform the **T** part.

This seems confusing, so let's look at the pattern behind this process. Below, you'll find some generic examples showing divisors of different lengths (expressed as an array of "d" digits) and a lengthy answer (represented by an array of "a" digits). *Each time you come to the "downward" step in a division problem of this sort, perform as many of these calculations as possible given the number of answer digits that are exposed by that point:*

Two-digit divisors:
NT: <u>dd</u> = aaaaaa<u>a</u>
  U: d<u>d</u> = aaaaaa<u>a</u>

Three-digit divisors:
NT: <u>dd</u>d = aaaaaa<u>a</u>

UT: d**dd** = aaaaaa**a**

  U: d**dd** = aaaaa**a**a

Four-digit divisors:

NT: **dd**dd = aaaaaa**a**

UT: d**dd**d = aaaaaa**a**

UT: d**dd**d = aaaaa**a**a

  U: dd**dd** = aaaa**a**aa

Do you see the pattern? After the **NT Calculation** (which always involves the two leftmost digits of the divisor and the rightmost answer digit), the **UT Calculations** follow a pattern that creeps toward the center. This applies to what you've learned so far in the following ways:

- With two-digit divisors, only use a **U Calculation** for the "downward" step. This is because there is no third divisor digit to perform the **T Calculation** with.
- With three-digit divisors, begin by using a **UT Calculation** for the "downward" step, because you now have the third divisor digit to work with.
- With three-digit divisors, once two answer digits are uncovered, use the sum of a **UT** and **U Calculation** for this step. This is because you're actually performing as many **UT Calculations** as you can, but there are only enough digits to begin the second one.

Once that's clear, let's keep going with the example. 6 (from **6**54) goes into the latest **Partial Dividend** (0) 0 times, so this is the next answer digit.

$91,560 \div 654 = 140$

Repeat the cycle again. Take this newfound answer digit (0) and perform an **NT Calculation** with it against 654.

6 (from <u>6</u>54) × 0 (from 14<u>0</u>) = <u>00</u>. Keep the entire answer (**N**).

5 (from 6<u>5</u>4) × 0 (from 14<u>0</u>) = <u>0</u>0. Keep the 0 (**T**).

0 (**N**) + 0 (**T**) = 0.

Subtract 0 (from the **NT Calculation**) from 0 (the latest **Partial Dividend**) to arrive at 0...

91,560 ÷ 654 = 140	Problem
<u>0</u>	Working Figures
0 (-0)	Partial Dividends

...which pairs with 6 from the above row to give you the next **Working Figure**, 06.

91,560 ÷ 654 = 140	Problem
0<u>6</u>	Working Figures
	Partial Dividends

From here, determine the next **UT** and **U Calculations. UT** first.

5 (from 6<u>5</u>4) × 0 (from 14<u>0</u>) = 0<u>0</u>. Keep the 0 (**U**).

4 (from 65<u>4</u>) × 0 (from 14<u>0</u>) = <u>0</u>0. Keep the 0 (**T**).

$$0 \ (\mathbf{U}) + 0 \ (\mathbf{T}) = 0.$$

Now do the **U Calculation**, using the second-to-last answer digit (remember the pattern) and the rightmost divisor digit:

4 (from 65**4**) × 4 (from 1**4**0) = 1**6**. Keep the 4 (**U**).

Add these two together:

$$0 \ (\mathbf{UT}) + 6 \ (\mathbf{U}) = 6.$$

Next, subtract this 6 (the sum of the **UT** and **U Calculations**) from the latest **Working Figure**. This gives you 0.

91,560 ÷ 654 = 140	Problem
06 (-6)	Working Figures
0	Partial Dividends

You crossed over the quotient/remainder line during the last step, but you're only now *entirely* on the right side of the line; at this point, you must approach things a bit differently. You will no longer divide the divisor into your **Partial Dividend**; instead, simply bring the untouched **Partial Dividend** up and to the right to become the leftmost part of the next **Working Figure**:

91,560 ÷ 654 = 140	Problem
<u>0</u>	Working Figures
0	Partial Dividends

Note that this was phrased "leftmost part" as opposed to "leftmost digit." You would do this no matter how many digits there are in the **Partial Dividends**. In this example, the **Partial Dividend** is 0, but if the **Partial Dividend** was 19, you'd bring 19 up in its entirety to become the leftmost part of the **Working Figure**.

Then bring down the 0 from 91,56<u>0</u> to complete the **Working Figure** (00):

91,560 ÷ 654 = 140	Problem
0<u>0</u>	Working Figures
	Partial Dividends

Multiply the rightmost digit of the answer (0 from 14<u>0</u>) by the rightmost digit of the divisor (4 from 65<u>4</u>) and subtract it from this **Working Figure**. Since 4 × 0 equals 0, you'll subtract 0 from 00. There is no remainder. The final answer is 140.

For now, take these remainder zone steps as law, but you'll understand them better when you move beyond three-digit divisors.

Let's review a few rules. To figure out where the quotient/ remainder line is, subtract 1 from the number of digits in the divisor, and place a little mark this many digits from the right side of your dividend. Once you've crossed entirely over this line and into the realm of remainders, you no longer divide the divisor into your **Partial Dividend**; instead, simply bring the untouched **Partial Dividend** up and to the right to become the leftmost part of the next **Working Figure**. Multiply the rightmost digit of the answer by the rightmost digit of the divisor and subtract it from this **Working Figure** to determine the remainder.

Do a few problems on your own. The answers are listed below. Write out each problem in its entirety as demonstrated above.

1. $67{,}140 \div 180$
2. $43{,}888 \div 211$
3. $109{,}557 \div 987$
4. $488{,}840 \div 880$

Answers:

1. 373
2. 208
3. 111
4. 555, remainder 440

Four-Digit Divisors (and Beyond)

Before moving beyond three-digit divisors, seriously ask yourself if you will ever need to use this skill; it's not practical, as situations rarely arise that would require you to perform this type of calculation precisely and without the aid of a calculator. However, if you're hooked and just can't help yourself, please feel free to read on.

Division problems with four-digit divisors are solved similarly to problems with three-digit divisors. The only difference is that for each extra answer digit you uncover, you must perform an additional **UT Calculation** during the downward step. This was touched upon when the pattern was illustrated earlier. Let's extend it.

Five-digit divisors:
NT: **d**dddd = aaaaaa<u>a</u>
UT: d<u>dd</u>dd = aaaaaa<u>a</u>
UT: dd<u>dd</u>d = aaaaa<u>a</u>a
UT: ddd<u>dd</u> = aaaa<u>a</u>aa
 U: dddd<u>d</u> = aaa<u>a</u>aaa

Six-digit divisors:
NT: **dd**dddd = aaaaaa<u>a</u>
UT: d<u>dd</u>ddd = aaaaaa<u>a</u>
UT: dd<u>dd</u>dd = aaaaa<u>a</u>a
UT: ddd<u>dd</u>d = aaaa<u>a</u>aa
UT: dddd<u>dd</u> = aaa<u>a</u>aaa
 U: ddddd<u>d</u> = aa<u>a</u>aaaa

Do you see how the pattern scales? Let's try a four-digit problem together: 7,006,652 ÷ 5,678. The divisor is purposely composed of adjacent digits (5,678) to help you keep track of which number you're working on.

First, drop the 7 down; it becomes the first **Partial Dividend**.

7,006,652 ÷ 5,678 =	Problem
	Working Figures
<u>7</u>	Partial Dividends

Take 5, the first digit of the divisor, and fit it into this first **Partial Dividend**. This gives you the first answer digit, 1.

7,006,652 ÷ 5,678 = <u>1</u>	Problem
	Working Figures
7	Partial Dividends

Perform the **NT Calculation**, using the first answer digit and the 5 and 6 from the divisor. This gives you 5, which you subtract from your **Partial Dividend**, 7, to arrive at 2.

7,006,652 ÷ 5,678 = 1	Problem
<u>2</u>	Working Figures
7 (-5)	Partial Dividends

Drop the second digit down from the top row, changing your first **Working Figure** from 2 to 20.

7,006,652 ÷ 5,678 = 1	Problem
2<u>0</u>	Working Figures
	Partial Dividends

Perform a **UT Calculation**, using the first answer digit and the 6 and 7 from the divisor. This gives you 6. Subtract that from your latest **Working Figure** to arrive at 14. Since there is only one answer digit to work with, you can't do any more complete or partial **UT Calculations**.

7,006,652 ÷ 5,678 = 1	Problem
20 (-6)	Working Figures
<u>14</u>	Partial Dividends

Repeat the cycle. 5 (from **5**,678) fits into 14 twice, so 2 is the next answer digit.

7,006,652 ÷ 5,678 = 1<u>2</u>	Problem
	Working Figures
14	Partial Dividends

Then perform an **NT Calculation**, using the newfound 2 and the 5 and 6 from the dividend. This gives you 11. 14 (the latest **Partial Dividend**) minus this 11 equals 3.

7,006,652 ÷ 5,678 = 12	Problem
$\underline{3}$	Working Figures
14 (-11)	Partial Dividends

Drop the next digit down from the top row to turn your latest **Working Figure** from 3 to 30.

7,006,652 ÷ 5,678 = 12	Problem
3$\underline{0}$	Working Figures
	Partial Dividends

Here's where things change because you're dealing with a four-digit divisor; you will perform multiple **UT Calculations**. First, perform one between the latest answer digit (2) and the 6 and 7 from the divisor; this gives you 3. Then perform one between the first answer digit (1) and the 7 and 8 from the divisor; this gives you 7. Add these answers together to get 10.

This may seem confusing, but if you keep the patterns in mind, it's not so bad. Remember, only address the pairings you can use using given the amount of the answer that's been uncovered. For instance, since only the "12" part of the answer is available, you'll perform only the following calculations:

NT: **5,**678 = **1**2
UT: 5,**67**8 = 1**2**
UT: 5,6**78** = **1**2

With these relationships in mind, let's continue with the problem. Next, subtract this 10 from the latest **Working Figure**, 30, to arrive at the next **Partial Dividend**, 20.

7,006,652 ÷ 5,678 = 12	Problem
30 (-10)	Working Figures
<u>20</u>	Partial Dividends

The cycle repeats; determine how many times 5 (from **5**,678) goes into this new **Partial Dividend**, 20. The answer seems to be 4, so you'll make this your next answer digit.

7,006,652 ÷ 5,678 = 12<u>4</u>	Problem
	Working Figures
20	Partial Dividends

Perform the **NT Calculation** using this new 4 with the 5 and 6 from 5,678. You end up with 22, which is too big (20 − 22 results in a negative number). Change your latest answer digit from 4 to 3 and try again.

7,006,652 ÷ 5,678 = 12<u>3</u>	Problem
	Working Figures
20	Partial Dividends

Perform the **NT Calculation**, using this new 3 against the 5 and 6 from 5,678, which gives you 16. Subtract this 16 from the latest **Partial Dividend**, 20, to arrive at 4.

7,006,652 ÷ 5,678 = 123	Problem
<u>4</u>	Working Figures
20 (-16)	Partial Dividends

Drop the 6 down from the top row to complete your latest **Working Figure**.

7,006,652 ÷ 5,678 = 123	Problem
4<u>6</u>	Working Figures
	Partial Dividends

Time for **UT** and **U Calculations**:

UT: 5,<u>67</u>8 = 12<u>3</u> Answer: 10
UT: 5,6<u>78</u> = 12<u>3</u> Answer: 5
 U: 5,67<u>8</u> = <u>1</u>23 Answer: 8

These add up to 23, which is subtracted from your last **Working Figure** (46). This gives you a **Partial Dividend** of 23.

7,006,652 ÷ 5,678 = 123	Problem
46 (-23)	Working Figures
<u>23</u>	Partial Dividends

5 (from **5**,678) goes into 23 about 4 times, so this is your final answer digit.

7,006,652 ÷ 5,678 = 123<u>4</u>	Problem
	Working Figures
23	Partial Dividends

NT: **5,6**78 = 1,23<u>4</u> gives you 22. Subtract this from your latest **Partial Dividend**. You've crossed into the remainder zone, but not all the way yet. Let's finish out this step.

7,006,652 ÷ 5,678 = 1234	Problem
<u>1</u>	Working Figures
23 (-22)	Partial Dividends

Drop the next untouched digit down from the top row to finish the **Working Figure**. This gives you 16.

7,006,652 ÷ 5,678 = 1234	Problem
16	Working Figures
	Partial Dividends

Time to get wild with **UT** and **U Calculations** again:

UT: 5,<u>67</u>8 = 1,23<u>4</u> Answer: 6
UT: 5,6<u>78</u> = 1,2<u>3</u>4 Answer: 3
U: 5,67<u>8</u> = 1,<u>2</u>34 Answer: 6

Added together, this equals 15. Subtract this from the latest **Working Figure** to arrive at the **Partial Dividend** 1.

7,006,652 ÷ 5,678 = 1234	Problem
16 (-15)	Working Figures
<u>1</u>	Partial Dividends

You are now entirely within the remainder zone, where the rules are different. If you don't remember how to locate the point where the quotient zone ends and the remainder zone begins, go back and review before continuing.

First, move your latest **Partial Dividend** up to begin your next **Working Figure** without doing any sort of **NT Calculation**.

380

7,006,652 ÷ 5,678 = 1234	Problem
1	Working Figures
	Partial Dividends

Drop the next untouched dividend digit down to complete it.

7,006,652 ÷ 5,678 = 1234	Problem
15	Working Figures
	Partial Dividends

You may recall from dealing with three-digit divisors that you determined the remainder by multiplying the rightmost digit of the answer by the rightmost digit of the divisor and subtracting the product from this **Working Figure**. In three-digit divisor examples, the line between quotient and remainder lies only two digits from the right-hand side of the dividend; therefore, by the time you're fully in the remainder zone, it's time to perform the aforementioned remainder calculation. With larger divisors, there's more room in the remainder zone, and so there are a few additional calculations to perform before this point.

You'll perform **UT** and **U Calculations**, but remove one of them each time you cycle through. The first one will be:

UT: 5,6<u>78</u> = 1,23<u>4</u> Answer: 11
U: 5,67<u>8</u> = 1,2<u>3</u>4 Answer: 4

This gives you 15, which you subtract from your latest **Working Figure**, 15, to arrive at 0.

7,006,652 ÷ 5,678 = 1234	Problem
15 (-15)	Working Figures
<u>0</u>	Partial Dividends

Move this **Partial Dividend** right up to become the first digit of the next **Working Figure**.

7,006,652 ÷ 5,678 = 1234	Problem
<u>0</u>	Working Figures
	Partial Dividends

Drop the next untouched digit down from the top row to complete the **Working Figure**.

7,006,652 ÷ 5,678 = 1234	Problem
0<u>2</u>	Working Figures
	Partial Dividends

Now, you perform one less calculation:

U: 5,67<u>8</u> = 1,23<u>4</u> Answer: 2. Subtract this from the **Working Figure** to arrive at 0.

7,006,652 ÷ 5,678 = 1234	Problem
12 (-12)	Working Figures
<u>0</u>	Partial Dividends

There's no remainder, but if there was, it would be this final **Partial Dividend**.

This seems difficult, but it isn't that hard once you get used to it. With some practice, you'll see that this method becomes quite easy, even for very length numbers.

Solve the following problems. You can write them out in their entirety at first. After you feel more comfortable with the method, try to work your way through the phases of **Mental Calculation** until you're able to see a problem written, save it to memory, step away, and solve it in your mind without further referencing it.

1. 10,018,701 ÷ 7,821
2. 2,412,128 ÷ 1,753
3. 3,148,677 ÷ 2,226

Answers:

1. 1,281
2. 1,376
3. 4141, remainder 1,113

Quick Review

1. The **Trachtenberg System** provides a method for dividing any number by any other number, regardless of size.

Review and Development: Section 7

Give yourself a pat on the back; you've gone above and beyond the call of duty and powered through some very difficult mathematical concepts. Let's review.

Review

1. **Merging Skills**: The principles of left-to-right addition/subtraction and **Familiar Numbers** can be combined into a three-step process for quickly solving two-number addition or subtraction problems. First the numbers are rounded to **Familiar Numbers** in such a way that you avoid **Overflows** whenever possible. Second, the problem is solved from left to right. Finally, account for any **Offset**.
2. **Trachtenberg Multiplication**: The **Trachtenberg System** provides a single method by which to multiply two large numbers together. This method scales in order to accommodate problems involving numbers with any number of digits.
3. **Trachtenberg Division**: The **Trachtenberg System** provides a method for dividing any number by any other number, regardless of size.

Development

You've endured quite a few *Review and Development* exercises by this point, so you know what to do here. Spend a few days developing the skills taught in *Section 7* while taking time to review and practice the skills you learned earlier in this book. By now, you should be intimately familiar with your study habits, time restrictions, etc., and know how to work at your own pace. After all, if you've made it this far, you have more than proven your work ethic and desire to succeed.

As this is the last section of the book, you may want to take the opportunity to construct your own development plan for any future skills you wish to learn, mathematical or otherwise. Create and execute a detailed day-by-day practice plan.

You've now learned every skill covered in *Master the Language of the Universe*. Before you go, there are a few things left to discuss, so please read on.

Conclusion

If this book was a graduation video, this would be the part where the screen fills with sentimental photos, transitioning softly from one to the next while Billy Joel croons *This is the Time to Remember*. It's been quite a journey.

Congratulations; you finished *Master the Language of the Universe*.

When you began reading this book, you were probably a little uncomfortable with mathematics. Along the way, you've encountered mentally taxing ideas, fought through countless hours of development exercises, and tackled frustrating, complex methodologies. Assuming you actually went through all of it and didn't skip ahead, you're now a shining example of how organized solution methods and hard work can transform a person. For this, you should be commended. You're a whiz. A shark. An ace. A mathemagician.

Before wrapping up, I'd like to briefly address a question that people tend to ask after learning these types of innovative mathematical methods: "Why isn't this taught in school?"

Why Isn't This Taught in School?

Fundamental, practical math skills are only taught to kids to the extent that they're needed as the foundation for more advanced skills. Pressured to move into more complex or abstract subjects like algebra, students pass beyond basic math well before practical application becomes quick and effortless.

For instance, when you first learned simple addition, how quickly could you compute $8 + 13$? When you were in school and found that you were able to perform this calculation in a few seconds, the teacher probably considered that a success and moved on, assuming that speed would come with a lifetime of continued practice.

For many, that speed never came.

As you moved through life, the world began handing you problems far more difficult than $8 + 13$. Having so quickly moved beyond basic addition years ago, what hope did you have to solve problems like $113 + 420 + 89$ with any grace?

People often ask whether the types of skills found in this book should be taught in schools. For the most part, the answer is no. Although one could argue that basic left-to-right addition/subtraction and **Familiar Numbers** could benefit the young, these skills are best suited for those who already possess a firm understanding of fundamental mathematics.

Some other topics in this book, such as **Guestimation, Probability, Options, Percentages,** and **Value Comparison** are touched upon in school; but like basic addition and subtraction, they are often abandoned before students are able to comfortably implement them in practical situations.

What about less familiar concepts, like math **Shortcuts** and the **Trachtenberg System**? Why aren't they taught in schools?

Math Shortcuts

If math **Shortcuts** are taught in school at all, they aren't part of the regular curriculum. The usefulness of **Shortcuts** are often underappreciated; they're instead seen as parlor tricks (and therefore stigmatized by their association with eccentric, magic-loving uncles).

Shortcuts are no substitute for the universally applicable core math skills taught in school; they're a luxury for those who understand fundamental mathematics and want to broaden their skill set. They cannot stand on their own and make no sense without the basic addition, subtraction, multiplication and division that fuels them.

The Trachtenberg System

As far as I can tell, the **Trachtenberg System** has never been taught in United States public schools. This isn't surprising, considering the lack of continuity between its methods for solving different problem types (especially the specific methods for multiplying by 5, 6, 7, 8, 9, 11, and 12). By contrast—though they're a poor choice for **Mental Calculation**—the traditional, classical multiplication methods taught in school are consistent and standardized (the same shared set of rules can be applied to any problem, regardless of size or complexity).

Using traditional math, the problems 128×19 and 128×12 would be solved the same way, but your approaches would be drastically different if using the **Trachtenberg System**; 128×12 would be solved using Trachtenberg's multiplication method for 12, whereas 128×19 would require the **Two-Finger Method**.

Once mastered, **Trachtenberg System** methods produce incredibly fast calculations compared to traditional math—especially when using only your mind. However, these methods are arguably more complex to learn and require specific approaches for each of the numerous types of problems.

Trachtenberg himself opened schools in his homeland to train young students, yet I feel strongly that these methods should be reserved for those familiar with classical arithmetic. Classical arithmetic provides a foundation of sound, logical mathematical principles that are not necessarily demonstrated by the **Trachtenberg System**.

If you or your child struggle with traditional math, please work on strengthening the fundamentals before attempting anything beyond *Section 1* of this book.

Beyond This Book

Should you independently research **Mental Calculation**, you'll discover quite a few systems not mentioned here, such as *Vedic Math* and the *Mental Abacus* method. In writing this book, these were systematically tested and ultimately disregarded in favor of the included methods. Still, you're encouraged to explore these and other methods if inclined to further pursue **Mental Calculation**.

To progress and stay sharp, you need to practice. None of the world's top chess players, pole-vaulters, or sharpshooters rely on natural ability alone; they all study and practice every day to stay at the top of their respective games. Regardless of any natural predispositions you may (or may not) have, you can become drastically more comfortable with math. Regardless of intellect, anyone can achieve some level of expertise in these skills. But like everyone else, you will need to work hard to maintain your abilities.

Anything worthwhile requires work. Don't let your effort go to waste by allowing everything you learn to dry-rot. I encourage you to craft a regular practice schedule and find ways to stick to it.

Thank You

Numbers are all around us, converging upon us, built into the science that seeks to explain our very existence and purpose. They are part of us, and we owe it to ourselves to go through life without fearing them. Enjoy.

Glossary

5-7 Approach   A method for the developing **Mental Calculation** abilities. It refers to five phases (of incrementally increasing complexity) and seven skills.

Addition Group   In any addition problem involving two numbers, the digits that occupy each number's respective "place"—such as the *tens* place or *hundreds* place—are together considered an **Addition Group**.

Addition Subtraction   A method for subtraction that involves splitting a subtraction problem into two smaller problems: a simpler subtraction problem and an addition problem.

Answer Digits Journey   A **Journey** used to remember the individual answer digits of a math problem.

Bubble Sort   A basic **Sorting** method.

Clue   See **Mnemonic Clue**.

Contextual Journey   A **Journey** location related to the information being memorized.

Direct Method	A **Trachtenberg System** method of speed multiplication that isn't explored in *Master the Language of the Universe* because it fails to scale gracefully. We instead explore the **Two-Finger Method**.
Double / Doubling	To **Double** a number is to multiply it by two.
Double Down	A multiplication **Shortcut**. To multiply any number by twenty: 1.) Double the number and 2.) multiply the product of the first step by ten. Expressed algebraically, this is $20n = 10(2n)$.
Dual-Offset Addition	A five-step addition method: 1.) Convert both of the problem's individual numbers to **Familiar Numbers**; 2.) Add the two **Familiar Numbers**; 3.) Figure out the **Offset** of each; 4.) Add the two **Offsets** together to arrive at a **Final Offset**; 5.) Account for the **Final Offset**.
Dual-Offset Subtraction	A four-step subtraction method: 1.) Round the two numbers in the same direction (the chosen direction doesn't matter), to arrive at two **Familiar Numbers**; 2.) Subtract the smaller **Familiar Number** from the larger one to arrive at the pre-**Offset** answer; 3.) Subtract the larger number's **Offset** from the smaller number's **Offset**; 4.) Subtract the **Final Offset** from the pre-**Offset** answer.
Duplicate Set	In **Probability**, this refers to all possible outcomes shared between two or more

possible outcome sets.

End-Merge Sort   A version of the **Merge Sort** (a basic
computer science **Sorting** method), adapted
for use by humans.

F2D Method   A method for determining exact decimal
values by memorizing a series of basic
fraction-to-decimal conversions and applying
the patterns they demonstrate.

Familiar Number   Any integer that—when divided by ten—
results in an integer.

Familiarizing   The act of rounding a number into an integer
that's divisible by ten.

**Fast Division
Method**   A **Trachtenberg System** method for precise
speed division.

Fast Method   See **Fast Division Method**.

Final Offset   The sum of all individual **Offsets** generated
while solving an addition or subtraction
problem.

Five-Ten Split   A multiplication **Shortcut**. To multiply any
number by five: 1.) Cut the number in half
and 2.) multiply by ten. Expressed
algebraically, this is $5n = 10(1/2\ n)$.

Fraction Group	A set of fractions that share the same numerator (e.g., 1/2, 1/3, 1/4, 1/5...)
FOIL	A multiplication **Shortcut**. To multiply any two two-digit numbers: 1.) Multiply the tens-place digits together and multiply the product by 100; 2.) Multiply the tens-place digit of the first number by the ones-place digit of the second number and multiply the product by ten; 3.) Multiply the ones-place digit of the first number by the tens-place digit of the second number and multiply the product by ten; 4.) Multiply the ones-place digit of the first number by the tens-place digit of the second number; and 5.) Add the products of steps one through four together.
Guestimation	The art of making numerical estimates based upon incomplete information.
Half-and-Half	A multiplication **Shortcut**. To divide any number by 4: 1.) Cut the number in half and 2.) cut it in half again. Expressed algebraically, this is $n \div 4 = (n/2)/2$.
Insertion Sort	A basic **Sorting** method.
Inside Digit	In a **Trachtenberg System** multiplication problem, this is the digit of the multiplier that is closest (of those being actively worked upon) to the multiplicand.
Inside Pair	In a **Trachtenberg System** multiplication

problem, this is a pairing involving **the Inside Digit** of the multiplier and the two rightmost digits of the multiplicand that are being acted upon.

Jigsaw	This is a multiplication **Shortcut** that only works for two two-digit numbers that share the same tens-place digit and have ones-place digits that add up to ten. 1.) Multiply the shared tens-place digit by "itself plus one." 2.) Multiply the two ones place digits. If this step results in a single-digit product, include a preceding 0; 3.) Finally, place the product of the first step to the left of the product of the second step.
Journey	A mental image of a real location, featuring multiple **Steps** or stops.
Journey Method	A mnemonic strategy in which information is stored along the individual **Steps** of imagined **Journeys**.
Limitation Set	When dealing with **Options**, this is the number of items from the original set that may be kept.
Manual Sorting	The act of **Sorting** objects without the aid of third-party tools.
Master Stack	In **Sorting** terminology, a **Master Stack** is an ordered, combined set comprising multiple **Preliminary Stacks**.

Master Value	When more than two numbers are involved in a subtraction problem, it follows that two or more of the numbers are being subtracted from a single (usually larger) number; this number is referred to as a **Master Value**.
Memory Development	The act of assessing and altering the methods you use to store memories so as to improve the efficiency and usefulness of your memory as a whole.
Mental Calculation	The act of performing addition, subtraction, multiplication, division, or other mathematical operations in one's mind, without the use of external tools.
Merge Sort	A basic **Sorting** method.
Middle Pair	In a **Trachtenberg System** multiplication problem, this is a pairing involving a multiplier digit and two interior multiplicand digits that are being acted upon.
Mnemonic Clue	A visual representation of a number or other piece of information. **Mnemonic Clues** can be people, creatures, objects, etc.
Mnemonics	Mental associations that function as memory aids, helping you to remember information more easily.
Mnemonic Scene	The imaginative events and interactions that

take place within a **Journey Step**.

Mnemonic Vocabulary	A personalized set of **Mnemonic Clues** used to represent a specific type of information.
Multi-Faceted Value Comparison	Comparing two or more options or values that require additional operations or unit conversion.
Nine to Ten and Down Again	A multiplication **Shortcut**. To multiply any number by nine: 1.) Multiply it by ten (or—phrased differently—"place a zero at the right side of the number"), and then 2.) subtract the original number from the product of the first step. Expressed algebraically, this is $9n = 10n-n$.
NT Calculation	A **Trachtenberg System** division action in which you multiply two numbers in a specific way: You keep the entire product of the first calculation and only the tens-place digit of the second.
Offset	When a number is **Familiarized**, an **Offset** is produced, reflecting the difference between the **Familiar Number** and the **Unfamiliar Number** from which it was converted.
One-Way Rounding Rule	A rule stating that—for simplicity's sake—all rounding in an addition or subtraction problem should occur in one direction (either up or down).

Options	The possible outcomes resulting from reordering or limiting a set of events or objects.
"Or" Subtotal	In discussing **Probability**, this is the answer before removing any **Duplicate Sets**.
Outside Digit	In a **Trachtenberg System** multiplication problem, this is the digit of the multiplier that is farthest (of those being actively worked upon) from the multiplicand.
Outside Pair	In a **Trachtenberg System** multiplication problem, this is a pairing involving the **Outside Digit** of the multiplier and the two leftmost digits of the multiplicand that are being acted upon.
Overflow	In addition, this refers to the "carry" or amount by which an **Addition Group's** sum exceeds ten.
Overflow Risk	In addition, this refers to any **Addition Group** whose sum is a number that—if affected by **Overflow** from a neighboring **Addition Group**—could result in **Overflow** itself.
Pair-Product	When using the **Trachtenberg System** for multiplication, this refers to a number resulting from a specific type of calculation wherein only certain digits of multiple products are utilized.

Partial Dividend	A number that exists on the bottom row of work in a **Trachtenberg System** division problem. **Partial Dividends** serve as numbers into which the first digit of the multiplier is divided in order to unveil answer digits.
Percentages	Human conceptual models are composed of individual objects or ideas that manifest as wholes or parts of wholes. **Percentages** are core to the language used to describe parts of a whole.
Pivot Point	In **Sorting** terminology, a **Pivot Point** is the item in a set that is located halfway between the first and last.
Practical Probability	The art of applying **Probability** principles— often somewhat loosely—to everyday situations.
Preliminary Stack	In **Sorting** terminology, a **Preliminary Stack** is an ordered set of items that contributes to the final **Master Stack**.
Pre-scanning	The act of quickly browsing the numbers involved in a math problem to preemptively identify potential problem areas.
Probability	The likelihood that an event will occur, assuming it is not known whether or not the event will *definitely* occur.

Quarter Pounder	A multiplication **Shortcut**. To multiply any number by 25: 1.) Multiply it by 100 and 2.) divide the product of the first step by four. Expressed algebraically, this is *25n = 100n/4*.
Quicksort	A basic **Sorting** method.
Short Splitting	When a number is made up of several digits (or pairs of digits) that are easily **Split**, you may choose to break the number up and **Split** the digits or pairs individually. For example, the number 826 is easily split using this method; 8 becomes 4, 2 becomes 1 and 6 becomes 3, resulting in 413.
Shortcuts	Tricks that allow you to solve certain types of math problems quickly and efficiently.
Simple Division Method	A **Trachtenberg System** method of speed division that isn't explored in *Master the Language of the Universe* because it fails to scale gracefully. We instead explore the **Fast Method**.
Simple Method	See **Simple Division Method**.
Simple Value Comparison	Comparing two or more options or values that require no conversion.
Simultaneous Events	When discussing **Probability**, this refers to two or more occurrences that take place at the same time.

Single-Digit Mnemonic Clue	An object or entity that represents a single number from zero to nine.
Single-Offset Addition	A four-step addition method: 1.) Convert the smaller of the problem's numbers to a **Familiar Number**; 2.) Add the larger number to the new **Familiar Number**; 3.) Figure out the **Offset**; 4.) Account for the **Offset**.
Sort / Sorting	The act of taking objects and placing them into an order based on specific criteria.
Split / Splitting	To **Split** a number is to divide it by two.
Splitsy-Doubly	A multiplication **Shortcut**. If at least one of two numbers that need to be multiplied happens to be a power of two (2, 4, 8, 16, 32, 64, 128, 256, 512...), work your way through the problem by continually cutting the "power of two" number in half while doubling the other. Repeat this **Split/Double** step until the problem is simple enough to solve.
Step	When utilizing the **Journey Method**, this refers to a single location within a **Journey**.
Subitization	The innate human ability to instantly recognize the number of objects in a small group (five or fewer objects) without counting them.
Subtotal	When adding three or more numbers, this

refers to the final count after all elevens have been removed.

Subtraction Group

In any subtraction problem involving two numbers, the digits that occupy each respective "place"—such as the tens place or hundreds place, for example—are together considered a **Subtraction Group**

Subtraction Subtotal

When more than two numbers are involved in a subtraction problem, it follows that two or more of the numbers are being subtracted from a single (usually larger) number (a **Master Value**). Add together all non-**Master Value** numbers first—this is called a **Subtraction Subtotal**—and subtract them from the **Master Value**.

Successive Events

When discussing **Probability**, this refers to two or more occurrences that take place one after another.

Teeny-Teeny

A multiplication **Shortcut** for problems with two numbers between ten and twenty. 1.) Add the larger number and the ones-place digit from the smaller number; 2.) Add a 0 to the end of your answer so far; 3.) Hold this combined number in your mind; 4.) Multiply the two ones-place digits of the two original numbers; 5.) Add the product of the fourth step to the number you stored in the third.

Temporary Stack

In **Sorting** terminology, a **Temporary Stack** is a subset of a **Master Stack** that is removed and focused upon for granular Sorting.

Tick	When performing **Trachtenberg System** addition upon three or more numbers, a **Tick** is an indication that a column's **Subtotal** has exceeded eleven and has therefore been decreased by eleven.
Tick Count	The number of **Ticks** that have been collected while performing **Trachtenberg System** addition upon three or more numbers.
Ticks Method	A **Trachtenberg System** method for quickly adding of three or more numbers.
Trachtenberg System	A set of specific math strategies devised by mathematician Jakow Trachtenberg. These strategies are well suited for **Mental Calculation** due to their avoidance of confusing aspects of math (such as excessive carrying and borrowing).
Two-and-Shift	A multiplication **Shortcut**. To divide any number by 5: 1.) Multiply the number by 2 and then 2.) move the decimal place one space to the left. Expressed algebraically, this is $n/5 = 0.1(2n)$.
Two-Finger Method	A **Trachtenberg System** method for multiplying large numbers.
U Calculation	A **Trachtenberg System** division action in which you multiply two numbers, keeping

only the "units place" (another term for "ones place") of the product.

Under Twenty Tables	This refers to the ability to immediately add or subtract any two numbers from zero to twenty. You should be able to do so before pursuing any further practical mathematical education.
Unfamiliar Number	An integer that—when divided by ten—does not result in an integer.
UT Calculation	A **Trachtenberg System** division action in which you multiply two pairs of numbers, keeping only the "units place" (another term for "ones place") of the first calculation's product and the "tens place" of the second.
Value Comparison	The act of comparing two or more objects or values.
Verbose Splitting	The act of dividing a number by two by breaking it down into its individual components, dividing each component by two, and adding all these results together. Take, for instance, 4,381 ÷ 2: Half of 4,000 is 2,000, plus half of 300, which is 150, equals 2,150. Half of 80 is 40, plus 2,150 is 2,190. Half of 1 is .5, plus 2,190 is 2,190.5.
Vocabulary	See **Mnemonic Vocabulary**.

Working Figure   A number that exists on the middle row of
                       Trachtenberg System division problem.
                       Working Figures serve as tools through
                       which **Partial Dividends** can be determined.

www.ingramcontent.com/pod-product-compliance
Lightning Source LLC
Chambersburg PA
CBHW060316200326
41519CB00011BA/1741